Forging Trust Communities

Forging Trust Communities

How Technology Changes Politics

IRENE S. WU

Johns Hopkins University Press
Baltimore

Printed in the United States of America on acid-free paper
9 8 7 6 5 4 3 2 1

Johns Hopkins University Press
2715 North Charles Street
Baltimore, Maryland 21218-4363
www.press.jhu.edu

Library of Congress Cataloging-in-Publication Data

Wu, Irene S.
Forging trust communities : how technology changes politics / Irene S. Wu.
pages cm
Includes bibliographical references and index.
ISBN 978-1-4214-1726-4 (pbk. : alk. paper)—ISBN 978-1-4214-1727-1 (electronic)—ISBN 1-4214-1726-X (pbk. : alk. paper)—ISBN 1-4214-1727-8 (electronic) 1. Political participation—Technological innovations. 2. Political participation—Computer networks. 3. Information technology—Political aspects. 4. Internet—Political aspects. I. Title.
JF799.5.W8 2015
303.48'3—dc23 2014039515

A catalog record for this book is available from the British Library.

Special discounts are available for bulk purchases of this book. For more information, please contact Special Sales at 410-516-6936 or specialsales@press.jhu.edu.

Johns Hopkins University Press uses environmentally friendly book materials, including recycled text paper that is composed of at least 30 percent post-consumer waste, whenever possible.

To my dearest husband, Joseph

CONTENTS

Acknowledgments xi

PART ONE: INFORMATION AND POLITICS: THEORY AND APPLICATION

1 Trust Communities from the Telegraph to the Internet: Information and Ideas as Capital and Ammunition 3

Activists Use the Latest Technology Available 5

Governments Use Technology to Define the Nation 6

The Link between Commercial Success and Political Usefulness 7

Sharing and Interaction Create Meaning within a Trust Community 8

Trust Communities Can Have Diverse Members 9

Information as Political Currency 10

The Trust Community as an Analytical Tool 11

Unpacking the Concept of "Trust Community" 12

2 Blogs, Wikis, and International Collective Action: The 2004 Boxing Day Tsunami 23

How This Case Came to Light 26

Failure of Government, Humanitarian, and Media Institutions 28

Why Individuals Came Together and How They Did It 29

Shock, Grief, and Anger 30

Frustration and the Impulse to Help 31

Creating the Blog and Wiki: Building a Repertoire 32

Making the Blog Easier to Use 33

Creating the Wiki 34

Moving the Wiki to a New Home 36

The Egalitarian Ethos: Philosophical Underpinnings of the Group 38

Blog and Wiki Effectiveness on the Ground 40
Did Volunteers Participate Again in Other Collective Actions? 42
The View through the Lens of a Trust Community 46
Creating a Common Identity 46
Building Trust 47
Social Capital 49
Network 49
Trust Community 49
Institution 50

PART TWO: NETWORK TECHNOLOGY CASE STUDIES

3 Activists Challenge Institutions with Information Technology Networks 55
China 1900: Protecting the Emperor with Public Telegrams 56
Philippines 2001: Phones, People Power, and Ousting of the President 58
Taiwan 1970s: Cable Television and the Rise of Democracy 61
Global 1990s: Banning Landmines with an Information Landslide 63
Egypt and Tunisia 2011: History, Social Media, and Revolution 67
Conclusion 71

4 Governments Shape Nations with Communications Technology 74
Infrastructure and National Identity 77
Canada 1927: Telephone Ties the Nation Together 77
Brazil 1900: Telegraph Reaches the Amazon 77
Discussion: Contrasting Brazil and Canada 79
Infrastructure, Economic Development, and National Security 80
China 1979: Telecom Growth versus State Self-Preservation 81
United States 1864: Telegraph as the Union's Secret Weapon 82
United States 1968: Internet Puts Commerce First 83
Information, Ideas, and National Security 84
USSR 1960: "InterNyet" Puts Information Control First 84
Russia 1880: Tsars Censor Printing 86
Information, Ideas, and Delivering Public Services 87
Global 1990: World Health Organization Learns from Crowds 89
United States 1960: Internal Revenue Service Gets a Computer 89
Information, Ideas, and National Identity 91

India 1987: Doordarshan TV's Ramayana *Recasts Politics* 91
UK 1938: BBC World Service Radio Embodies Empire 93
Qatar 1996: Al-Jazeera Satellite TV Raises Qatar's Profile 95
Discussion: BBC, Doordarshan, and Al-Jazeera 96
Conclusion 97

PART THREE: TRUST COMMUNITIES IN POLITICS

5 Technology + Trust = Political Influence 103
Trust Communities—Opportunities for Individuals and Institutions 105
The Role of Capitalism 108
Engagement, Participation, and Interactivity 109
Trust Communities and Diversity 111
Information and Ideas as a Source of Power 113
Trust Community as an Analytical Lens 114
Future Research 116
Conclusion 119

Epilogue: Using Technology to Lead: A Note to Activists, Businesses, and Governments 123
For the Activist: On Information as a Tool for Change 123
For Businesses: Your Customers' Trust Communities 128
For Governments: How Information and Ideas Shape the Nation 130

Notes 137
References 145
Index 153

ACKNOWLEDGMENTS

There are many things I love about academic life. One is asking people to share their life stories, and another is sitting in a library with a pile of books to read. This book afforded me the opportunity to do both, and for that I am grateful.

In 2005 I presented a glimmer of an idea at an International Research and Exchange Board (IREX) program. In 2007-2008 I took leave from government service to dive into the research while I was the Yahoo! Fellow at Georgetown University's School of Foreign Service, a program ably led by John Kline, Marjory Blumenthal, Casimir Yost, and James Seevers. For several years, my colleagues' enthusiasm maintained the project's momentum at weekly meetings of Georgetown's Book Lab, a club of scholars writing books led by Carole Sargent. Students in Georgetown's Communications, Culture, and Technology Department regularly questioned my reasoning. Steve Leu and Elizaveta Chuykova were enthusiastic research assistants. And in the background, my family and friends were unswervingly confident, even when the early explanations of what I was about did not match the clarity of the book today. My thanks to all of them.

To the many *TsunamiHelp* volunteers I interviewed, thank you for sharing your time and experience. Your insights helped connect the present to the past.

Several colleagues provided direct comments on the book manuscript, in whole or in part, improving it immeasurably. I thank especially Nanette Levinson, Sandra Braman, J. P. Singh, and Jeffrey Hart. Peter Rutland, Amit Schetjer, Richard Taylor, Sascha Meinrath, Christopher Smith, Leah Shapiro, and Matt Lussenhop invited me to their respective schools and institutions to present. The lively discussion from audiences ranging from undergraduates to seasoned senior officials was fun and enlightening.

In the early stages of this research, numerous colleagues spent hours over coffee, tea, and other refreshment engaging ideas high and low. They include Carolina Matos, Pablo Sotero, Carlos Lins da Silva, Tara Nair, Anjali Monteiro, K. T. Jayasakar, Seem Khanwalker, Alan D'Souza, A. F. Mathew, Rehana Ghiadally, Farida Umrani, S. C. Bhatnagar, Karita Das Gupta, Ang Peng Hwa, Adam Lifshey, Brian McCann, Vivaldo Santos, Tristan James Mabry, and Michael Fereira. While they may recognize little in the book now of what we discussed then, I remember their gracious hospitality and open minds.

As an author situated between academia and government, I am grateful for the welcoming research community fostered by numerous professional organizations. In particular I benefit from the camaraderie at the American Political Science Association, the Information Technology and Politics Section and the Practicing Politics Working Group; the International Studies Association's International Communications Section; the Telecommunications Policy Research Conference; and Politics Web 2.0 at the University of London Royal Holloway.

As this book is published, I serve as an analyst at the Federal Communications Commission. Many of my FCC colleagues have been kind and supportive of my extracurricular research. However, of course, I take full responsibility for it. The book does not reflect the views of the agency staff or the commissioners.

Finally, thanks to my editors Suzanne Flinchbaugh, Kelley Squazzo, and Catherine Goldstead and their colleagues at Johns Hopkins University Press for guiding this book to completion.

PART ONE

Information and Politics

Theory and Application

CHAPTER I

Trust Communities from the Telegraph to the Internet

Information and Ideas as Capital and Ammunition

ALTHOUGH THE INTERNET is still new, using technology to communicate is as old as cave paintings. When the telegraph was invented, a message that once took forty days to travel from London to Hong Kong suddenly could be delivered in a few hours, sometimes within minutes.[1]

We have seen this kind of time and space compression before; however, the Internet and mobile phones bring something new to politics. Chinese protestors against Japan organize by texting with mobile phones. Terrorist groups pour out their messages and recruit new members on websites. This is the new public square.

The same technology is also enriching individual lives. Online support groups give strength to people who suffer, whether from disease or discrimination. Online, even a very specific interest can attract a critical mass of people. Individuals can explore a latent identity—Celtic speakers abroad—or an unusual hobby—growing African violets. This is the new private sphere.

In the past, networks of people might communicate by letters or word of mouth. Now they use email and satellite television. Networks are often loosely connected. Sometimes, however, members of a network closely interact, reciprocate favors, and build trust. This trust enables them to cooperate. When this happens, the network becomes a *community,* and that *trust* is a major asset. These *trust communities* can be small—neighbors petitioning the city council for a new traffic light—or large—public health authorities with a message about this year's flu reaching across national and local borders to public organizations and private corporations, to groups and individuals.

Past studies of the Internet and politics split two ways. Some see the Internet as changing little: governments exercise military and economic control over protests online just as they do over protests in the streets. Before the Internet, politicians raised election campaign donations by letter or fax; now this is accomplished by email and texting. This is simply technology extending the old politics. However, some believe the Internet changes everything; more information is available, greater transparency is unavoidable, and new institutions arise to replace the old. Actually, both are occurring. Sometimes new communication service technology not only extends political activity but also transforms it.[2]

For individuals, a wave of new information and ideas can either strengthen their connection with old ties or put them in touch with new ones. Perhaps an onslaught of the foreign will mean that someone clings closer to the village trust community of families, neighbors, and ancestors. On the other hand, exposure to fresh ideas may mean that someone reaches out to new trust communities and latent identities find expression. New technology can change individuals' sense of themselves relative to the world.

For institutions, new technologies also change horizons. Institutions themselves are networks of people—people who are part of the institution's inner workings, people who are participants in the institution's causes, people who watch the institution's work, people who oppose the institution's work. Whether through an onslaught of the foreign or an introduction of fresh ideas, a new wave of information changes the choices available to the people in the institution's network. Red Cross supporters have more information about other international humanitarian organizations; progressive political parties in one part of the world learn about the environmental agendas in another part; citizens in one country learn about the social safety nets of citizens in another country. Old institutions compete with new ideas. In the face of new sources of information or alternative interpretations of meaning, the old institutions can either adapt or face extinction.

New communications technology transforms politics by increasing opportunities for people to build reciprocity and trust, by opening new possibilities to create community and build social capital, and by establishing the necessary conditions for collective action. It is not the technology that changes politics; it is the enhanced relationships among people—enabled by technology—that changes politics. The creation, distribution, and consumption of information and ideas can be the core of political life in this kind of trust community. Under these circumstances, since technology affects how information and ideas are distributed, it also influences how political power is distributed. Both political activists and governments use the latest communications technology—whether the telegraph, satellite television, or social media—toward their own ends. By placing the Internet in historical context, this book unpacks the elements that connect technological innovation with political change. What may be new about twenty-first-century technology is that the richness of sharing online may come to rival face-to-face communications.

Activists Use the Latest Technology Available

An example of how new technology extends old political practices comes from China in the last century. In 1900, telegrams were used in China to mobilize sentiment against the Empress Dowager Cixi, who had placed the reformist Emperor Guangxu under house arrest. In China, for centuries, when people felt wronged, they traveled to the capital and presented petitions to the emperor that aired their grievances. In the new electronic communications age of the telegraph, people protested Empress Dowager Cixi's actions by sending telegrams, supported by signatures, to newspapers. These "public telegrams" were then printed in newspapers and circulated, creating public pressure on the empress dowager not to remove him. Activists such as Jing Yuanshan, who was also chief of the Shanghai

Telegraph Administration, sent telegrams to newspapers around China and around the world. This triggered more telegrams in support of the emperor. Some telegrams originated within China and others from abroad. These telegrams, one of which was signed by more than 1,200 people, appealed to the empress dowager not to depose Emperor Guangxu. With the advent of public telegrams, instead of taking weeks to get information, newspapers were able to gather news in two or three days. Newspapers that adapted to the public telegrams grew in readership; those that did not saw circulation fall. In the end, Empress Dowager Cixi, surprised by the vehemence of public opinion, was not able to replace Emperor Guangxu, but the activist Jing Yuanshan was later arrested in Macau.[3]

The protection of Emperor Guangxu was neither the first nor the last time people used public telegrams to express their opinions to the Qing government. While no formal institution emerged from these public telegram campaigns, the advent of the telegram did enable a new trust community that followed in an old Chinese political practice.

Governments Use Technology to Define the Nation

Governments use communications to govern. Their objectives are to enhance national identity, preserve national security, and promote economic development. They thus establish institutions and frameworks to influence the production and distribution of information, though its interpretation, of course, remains beyond their control. A recent example of a government expanding its influence by extending a communications service is Qatar and the satellite news service Al-Jazeera. In the 1990s, the leader of Qatar, Sheikh Hamad, took clear and decisive actions to distinguish Qatar from its fellow Gulf States, hosting major international political and sports events and making Qatar an education leader. In 1996, the Emir launched Al-Jazeera news service and revolutionized the news in the Middle East. For the first time, television audiences had access to professional, in-depth coverage of political issues. It also succeeded in giving

Qatar a global profile and regional leverage that it did not have before. Qatar's influence as a country benefits from the success of Al-Jazeera.[4]

The nation is one type of trust community. As Benedict Anderson argued in *Imagined Communities,* a nation is held together by a common notion of being tied together, in his analysis, by the everyday chronicling of the newspaper. The system of newspapers is a trust community of newsstands, reporters and their sources, readers and their friends, advertisers and their customers, and printers and their suppliers. A nation's trust community consists of citizens and would-be citizens, politicians and their rivals, popular culture leaders and their critics, business and civil society organizations—all those who feel some relationship with the nation. Managing ideas and information to create a strong trust community is one of the basic functions of the state. At the most fundamental level, the practical manifestation of the state's management of these relationships is in its policy toward information and communications.

The Link between Commercial Success and Political Usefulness

The commercialization of communications technology is the foundation of that technology's usefulness as a political tool. Anderson identified the innovations in printing presses and development of commercial newspapers as instrumental in constructing a national identity.[5] The quickest way for a new idea to take hold is for it to travel across communication infrastructure built into every household and firm, a service that is popular and commercially successful and that links people who have a common cause or shared identity. In other words, every household may buy a television in order to watch soap operas, but once the television is an established service in everyday life, television programming can convey politically relevant messages as well. The principles of supply and demand are critical to understanding the commercial popularization of a communications technology and also apply to the information and ideas the technology carries. The technology becomes popular only if customers are willing to buy it and use it.

In 1865, when Paraguay invaded Brazil's southern Matto Grosso, it took the capital, Rio, six weeks to hear about it. In 1889, when the monarch in Rio was overthrown and the country declared a republic, it was a month before word reached the residents of Matto Grosso. To speed up communications, the new republic of Brazil resolved to build a telegraph system through its interior states to link them to the rest of the nation. Starting in 1890, Candido Mariano da Silva Rondon led a series of military units and commissions to establish the first telegraph network across the Amazon. But in 1921, six years after the line was inaugurated, more than 80 percent of all telegraph messages were government communications. As a tool of the state, it had some results, but the extended benefit of encouraging development had not materialized.[6] The Amazon lands were not incorporated under the center's control, development did not occur, and indigenous people were not assimilated.

This illustrates the intersection of politics and economics, or the political economy, of how trust communities develop and the use of information as a source of political power. Entrepreneurial companies, motivated by profit, cater to customer demand; this dynamic often more quickly results in the popularization of a technology than do government offices deploying a technology in pursuit of a policy objective. Even if governments succeed in distributing a technology, they then face the challenge of encouraging people to adopt it.

Sharing and Interaction Create Meaning within a Trust Community

Trust communities develop when there are plenty of opportunities to interact and reciprocate. For example, after President Estrada came into office in the Philippines in 2001, he sought to dampen criticism of his administration by bribing and threatening news organizations. However, people were aware that the reporting did not align with their experiences. They began turning to alternative sources of information, often distributed through short-message-service texts on cell phones, for example. By the time a surfeit of suspicion mounted around Estrada's corruption, people used these alternative networks to stay informed, mobilize demonstrations, and eventually force the ouster of Estrada.

The people who produce information cannot control the interpretations actually taken by those who receive it. With Internet applications such as social media, it is easy for audiences to express their views. In older communications media, audiences also have their own views; however, identifying and understanding them requires more work and attention.

Trust Communities Can Have Diverse Members

On December 26, 2004, a major tsunami hit several countries in South and Southeast Asia. The magnitude of the disaster across the entire region was of a scale that had not been experienced for many years. Among the range of responses that developed, one was an entirely online effort. A small group of bloggers in India spearheaded the creation of a blog (http://tsunamihelp.blogspot.com), later transformed into a wiki (www.tsunamihelp.info/wiki/), that for a few weeks became a major international clearinghouse for people around the world interested in providing assistance to the victims of the tsunami. The blog and wiki arose because there was a void not filled by existing national and international institutions. Governments were slow to report; the traditional news media had yet to reach the area. Around the world, people who wanted accurate news or who wanted to help found that their governments and other agencies were not responding fast enough with good information.[7] People from around the world started sending messages on what they knew, what they had, who needed what, and what was needed where.[8] The effort was so successful that within a week of its creation, the blog *TsunamiHelp* was among the top ten humanitarian websites visited in the world, just behind the United Nations, Reuters AlertNet, and the International Federation of Red Cross and Red Crescent.

The *TsunamiHelp* volunteers were largely strangers to each other before the event. Most never met face-to-face. They were different ages, nationalities, professions, and backgrounds, yet they managed to cooperate intensely to great effect. Technology-enabled trust communities are often diverse by the usual standards of social analysis such as ethnicity, class, or religion. These trust communities are easy to join and easy to leave; people stay because they are committed to the community's cause or identity.

Information as Political Currency

The *TsunamiHelp* wiki and blog traded in information, not aid; and its information cache was the source of its political significance. Another example of how information can be used for direct political change is the International Campaign to Ban Landmines. When Nobel Peace Prize laureate Jody Williams describes how she, as coordinator of the International Campaign to Ban Landmines (ICBL), and the member nongovernment organizations (NGOs) were able to launch a coalition in 1992 and only five years later have a treaty signed by 122 countries, she emphasizes the importance of clear and consistent communications among the trust community. Phone, fax, and email were essential to communications internal to the coalition; face-to-face communications were critical between the coalition and government and military representations.[9] The ICBL coalition grasped that technology gave it an advantage, even over nation states. Indeed, Williams recollects, "the ICBL has often learned of developments relating to the ban movement before governments became aware of them. This has made the ICBL a focal point of information for governments and NGOs alike. Its role as an information center also helped build confidence between governments and the ICBL."[10] In other words, the ICBL trust community was so successful in bringing together information that it became as significant as national governments in this arena.

Technology-enabled trust communities often trade in information and ideas, the way businesses use currency and militaries use ammunition. It is not just activists who use information to establish their power base; governments do as well.

The Trust Community as an Analytical Tool

If information is a source of political power, how does that affect our analysis of society? As a guide, we can look to the industrial revolution. Technical inventions that enabled the transmission of electricity, enabling a single worker to harness the power equivalent to several horses, triggered the industrial revolution.[11] Observing the factories newly sprung around him, Karl Marx argued that society was divided between the capitalists, the people who owned the factories, and the wage laborers who worked in them. Over time, we have refined his distinction between the capitalists and the workers to our current notion of socioeconomic classes distinguished by level of income and type of work. Using class to understand society is distinct, for example, from using differences in religion, language, or family ties. From this recognition of economics as a fundamental organizing force in society, comes our understanding of it as a basis of political power for the nation state. Just as class as an analytical tool enables us to connect individuals with dynamics in the economy, so also do trust communities give us a tool to understand how individuals connect with changing information and ideas in the world. These trust communities are not just the physical networks of the communications infrastructure but are the network of human beings connected by technology and held together by some common idea—an identity, a project, or a hope.[12] These can be an identity suppressed by history, a project to change a society, a hope to create a new community.

In Taiwan from the 1950s on, political opposition was underground. In the mid-1970s cable operators in greater numbers illegally installed videocassette recorders, coaxial cable, and transmission equipment.[13] Cable television, collectively known as the "fourth channel," boomed because people wanted more entertainment, but in addition it provided news from the opposition point of view.[14] In

1990 the opposition Democratic Progressive Party (DPP) announced formation of Taiwan Democratic Cable Television Association, a group of about fifty cable systems that broadcast news and information from the opposition perspective, in part to seek legalization of cable TV.[15] In 1993 the ruling Nationalist Party, faced with pressure from two opposition parties and foreign investors, passed the Cable Law, which in effect legalized the expression of opposition views on television.[16] Opposition politics thus moved from underground, to illegal cable television, to full legal participation in a democratic system.

Prior to the industrial revolution, kingdoms built their power bases on the control of land. With the coming of the industrial revolution, ownership of land was no longer the main route to status, wealth, and power. Capital was another option. As Karl Marx argued, the essential change that the industrial revolution brought to the organization of society was the distinction between those who own capital and those who do not. Those who own capital, profit. Those who do not must sell their labor.[17] Now it is not just capital and land that are possible paths to wealth and power. Information is a third path, another basis of power that divides society among those who have it, control it, and understand it, and those who are at its mercy.

Unpacking the Concept of "Trust Community"

A few conceptual tools, simply explained, will help clarify comparisons among the nearly twenty case studies in this book. To begin, calling a community a "trust community" brings special attention to the relationships among its members. This builds on a tradition of study that seeks to understand what binds communities together and what dynamics enable them to work together in their collective interest even in ways that may be in tension with individuals' self-interest. The social glue that holds people together can be thought of as different strengths: identity, trust, and social capital. In a community, each of these qualities can be transformed in scale and character when new communications technologies are introduced. The puzzle is in understanding when it does or does not happen. Furthermore, when exploring the potential for the idea of a "trust community" it is useful to put it in the context of more widely used terms—network, community, and institution. Each of these terms—identity, trust, and social capital—has its own history, usage, and relationship to the social glues.

Identity is a person's sense of self and may motivate a person's actions. In his work on why people cooperate, Tom Tyler shows that there are two aspects of identity—social and emotional—that explain why an individual may cooperate in the interest of the community rather than acting selfishly. Both of these aspects of identity rest on a fundamental need of people to maintain a favorable and positive sense of self. *Social identity* is how people define their status through their membership in a group. The more strongly a person identifies with the group, the more completely he or she merges individual goals with the group's goals. Group membership also gives individuals a sense of pride and an expectation of respect from other members, both of which motivate people to cooperate.[18] Seen from a different angle, people will avoid adopting signs that they belong to groups that are not respected—such as carrying a book by opposition politicians that are vilified by society—until a time comes when that opposition group gain respect.[19] *Emotional identity* is another important aspect that explains people's willingness to cooperate. Psychologists show that people have a fundamental need to have attachments to others and will act to maintain positive, significant personal relationships.[20]

Trust has many facets; the aspect most relevant to this study is trust that enables cooperation. Why is it that people trust each other enough to cooperate, when acting individually might be in their self-interest? Behavioral social scientists like Elinor Ostrom have conducted experiments that show trust can be the result of repeated interaction. For example, a series of communications can lead one partner to believe the other partner can be relied on to reciprocate.

When such series multiply, people in a network begin to form expectations about others' behavior. They trust each other, and then it is easier for them to cooperate.[21]

Social capital makes it easier for members of a community to take action together. It includes trust, norms, and networks, as Robert Putnam puts it in his works on collective action. *Trust* is the expectation that others will reciprocate. *Norms* identify when that reciprocity can be expected. *Networks of civic engagement* are those intense interactions across society in groups like neighborhood associations, sports leagues, and political parties. The boundaries of these networks define the scope of possible action.[22] In Putnam's analysis there are two kinds of social capital—bonding social capital among people who are similar, and bridging social capital among people who are not similar. It is bridging social capital that is the hardest to create and the most valuable when it comes to cooperation.[23]

Brought together, these concepts of identity, trust, and social capital are kinds of glue that hold people together and enable them to work collectively in the group's best interests (fig. 1.1). Individuals are motivated when they identify with a group and trust the other group members. Groups are successful at collective action when there is trust among members and a fund of social capital.

There are several ways to describe groupings of people in society; three that are relevant to this study are networks, communities, and institutions (fig. 1.2).

A *network* is a structure of links and nodes. This study focuses on networks of human beings, as compared to networks of computer machines, for example. Individuals are nodes and are linked together by communication—whether face-to-face, email, or the latest social

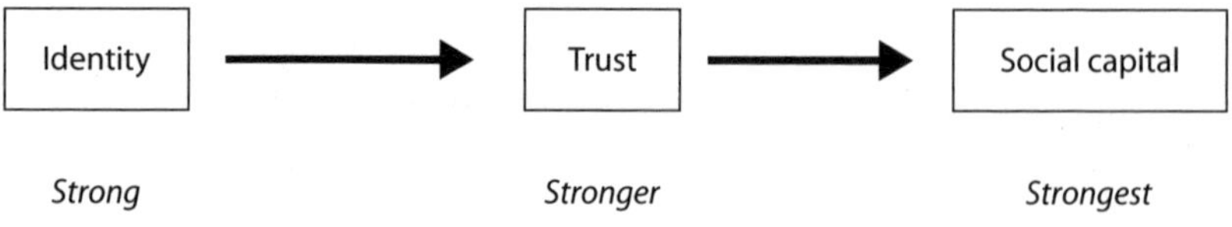

FIGURE 1.1. Social glue.

media tool. Milton Mueller in his study of networks and states defines network organizations that, while loosely tied together, still have definite boundaries; they are consciously constructed by their members and leverage expectations of reciprocity. He also underscores that a distinguishing characteristic of network organizations is that they are nonhierarchical.[24] One can think of network organizations as networks nested within larger networks.

Community is a group of people bound together by some common characteristic. It may be as simple as a common geographic home or a common place of work. It may be a shared interest like a hobby, or a shared cause like improving the environment. A community has generally accepted values, some level of homogeneity, and a specific size and composition.[25]

Institutions are communities with specific rules that govern the repetitive, structured interactions among its members, enabling them to act more effectively as a group. These rules exist within families, neighborhoods, markets, firms, sports leagues, churches, private associations, and government at all scales. These rules govern choices, and there are consequences for the chooser and others. Institutions that endure have clearly defined boundaries, clear cost and benefit tradeoffs, collective choice arrangements, accountability, graduated sanctions, and conflict resolution mechanisms.[26]

The idea of "trust community" joins the ideas of network and of community as a social group with the capacity to take collective action but without the formal rules and enforcement usually associated with institutions. Some institutions may be trust communities, especially if communication is a major aspect of their work, but many trust communities will not have the clearly defined membership or

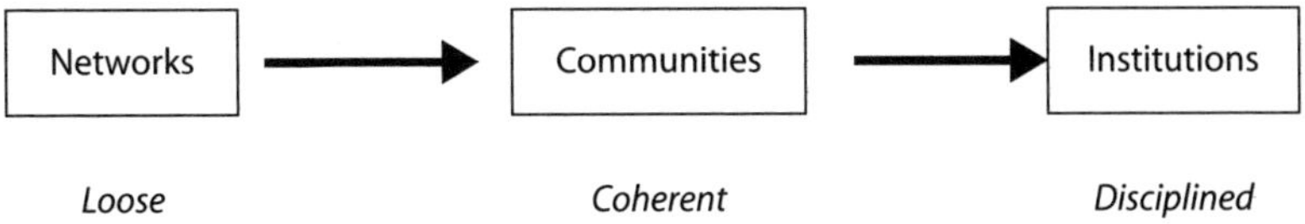

FIGURE 1.2. Organized social groups.

	Identity	Trust	Social Capital
Network	Yes	Maybe	No
Community	Yes	Yes	Maybe
Institution	Yes	Yes	Yes

FIGURE 1.3. Social groups and social glue.

boundaries typically associated with an institution. In figure 1.3, networks are held together by identity; communities, by identity and trust; and institutions, by identity, trust, and social capital. Trust communities are highlighted in gray.

The cases in this book are about activists—those groups making the transition from a network to a trust community—and governments—those institutions seeking to maintain the commitment of their members in the face of competing communities and networks.

Every major innovation in communications technology transforms the kinds of human networks that are possible and opens up the possibility of new and different kinds of communities, which in turn may give rise to new and different institutions. The possibility of transformation, however, does not mean that it is inevitable. Furthermore, new communications technologies can reinforce old networks, communities, and institutions. This book covers nearly twenty cases from a dozen countries and regions and from the nineteenth to the twenty-first centuries that demonstrate that the transformation of politics by communications technology is not a new phenomenon. The cases were chosen to illustrate how a trust community analysis can be applied to a range of communications technologies—telephone, television, and Internet—in a range of countries. A full list of the cases in this book is summarized in table 1.1.

These historical cases illuminate one of the most interesting political questions today: Does the Internet magnify the political power of the state or fundamentally challenge it? Neither, conclusively, I argue. Rather, the Internet shows that information is an emerging basis of political power. The technological innovations of the late twentieth

TABLE 1.1 Summary of Trust Communities—Historical Case Studies

Date	*Country*	*Case*	*Community members*	*Technology*	*Interpreters of info and ideas*	*Network scope*
1860	US	Civil War and telegraph	government, military, industry	telegraph	military	national
1900	China	Public telegrams	business and civil society leaders	telegraph, newspapers	signatories to telegrams, newspaper readers, imperial court	dissenters in China and Chinese diaspora worldwide
1917	Russia	Control over the newspaper *Word* (later *Isvestia*)	readers, publishers, censors, political leaders	improved printing press technology	readers, reporters, censors	Russia, later Soviet Union
1920s	Canada	Golden Jubilee	government, telegraph and telephone companies, radio audience	telephone, telegraph, radio networks	people of Canada, government of Canada, telephone companies	Canada
1930	Brazil	Telegraph to the Amazon	military, government	telegraph	military, government	Brazil
1930	Great Britain	BBC World Service and domestic service	reporters, audience	radio, television	reporters, audience, government politicians	global
1960	US	ARPANET	researchers, governments, businesses, public	computer, email	users	US, then global

(*continued*)

TABLE 1.1 *(continued)*

Date	*Country*	*Case*	*Community members*	*Technology*	*Interpreters of info and ideas*	*Network scope*
1960s	USSR	Computer network control	researchers, government	computer	government prevented it from happening	USSR
1980s forward	Taiwan	DPP and cable television	political opposition and citizens who support them	cable television news, newspapers and magazines	cable television viewers, opposition politicians, Nationalist government in power	within Taiwan
1980s forward	China	Telecom development	government, telecom companies, consumers	telecom	users	China
1980s	India	Hindu teleserials	television producers, public broadcaster, audience, politicians	television	audience, producers, politicians	India
1990	Global	International Campaign to Ban Landmines	nongovernment organizations, humanitarian agencies	email, fax, phone, face-to-face meetings	victims and potential victims, governments, military	global
1990s forward	Qatar	Al-Jazeera television	news reporters, audience, governments	television, Internet	governments, reporters, audience	global
1990s forward	US	Internal Revenue Service	taxpayers	Internet, personal computers	taxpayers	US
1990s forward	Global	World Health Organization	health professionals, general public	Internet	health professionals, general public	global

2000s	Pakistan Korea China Burma	website filtering in Pakistan, Korea, China, Burma	content producers, viewers, governments, political opposition	Internet, newspapers	targets of filtering, governments that filter, outside observers	national
2001	Philippines	Ouster of Estrada	political opposition, citizens in the street	mobile phones and SMS, television and radio reporting, newspaper reporting	citizens in the street, civil society organizations	Philippines
2004-5	Global	*TsunamiHelp*	volunteers	Internet—for publicity and recruiting volunteers, radio and newspapers	volunteers, humanitarian organizations, people affected and their friends and families, journalists looking for information	global
2011	Egypt Tunisia	Popular uprisings in Egypt and Tunisia resulting in ouster of political leaders	citizens, political leaders	social media, satellite television	citizens, political leaders, international community	Middle East, global

and early twenty-first century make information as significant as military and economic power were in the past.[27] First, this book shows how revolutionaries and activists have used new communications technologies to challenge the state. Second, this book demonstrates how, historically, governments have sought to control communications networks, especially as new technologies emerged, in order to extend their own power. Finally, the book concludes with ideas about how information as a power base affects the work of activists, the decisions of government policymakers, and our theoretical understanding of politics and technology.

Chapter 2 investigates *TsunamiHelp,* showing how the Internet enabled individuals to compete with international organizations in providing information to the public. Historically, however, every new communications technology enables some dissident or marginalized groups somewhere to link together in trust communities as never before.

Chapter 3 chronicles the history of earlier communications technologies as used by political activists, opponents to the state. As mentioned earlier, in the 1980s, Taiwan's political opposition mobilized people to their cause through cable television. In the 1990s, nongovernment organizations organized a campaign with phone, faxes, and email that culminated in a 1997 treaty to ban landmines. In 2001, Filipinos erupted in protest against a corrupt President Estrada, using mobile phones and SMS (short messaging service) as personal broadcast stations, eventually leading to his resignation. Similar protests in 2011 Egypt and Tunisia ousted their respective leaders. However, we can reach further back in history. At the beginning of the twentieth century, protestors in China used the telegram to organize against the Qing dynasty. The common link between the contemporary and the historical is that the technology enabled the creation of competing channels of information that offered participants an alternative view of the politics.

In each of these cases, a technology new at the time facilitated the creation of a new trust community whose members began exchanging information and ideas. The people in these trust communities developed new tools and techniques for collecting, verifying, and distributing ideas. Very often these trust communities were able to continue applying these tools and techniques to a variety of new situations, laying the groundwork for new institutions. Some of these new institutions, like the political opposition in Taiwan and the International Campaign to Ban Landmines are now major players in their respective political arenas.

Chapter 4 examines historically government's efforts to control communications technology innovations as they have emerged—sometimes successfully, sometimes not. Governments strenuously attempt to magnify their political power by extending communications services. The rationale is if a government is successful in establishing a complex, comprehensive communications system, then its own messages to the people will be readily transmitted. In practice, this is not necessarily the case, thus underscoring that communicating is an exchange, not a one-way transmission. In the case of Canada, political pressure to confirm a unique national identity catalyzed the development of a national telephone network more quickly than would have been expected solely on a commercial basis. However, in Brazil pure political motive and the complete lack of commercial basis undermined the national telegraph project. In the era of satellite television, competition pushed government broadcaster Doordarshan to shift away from its unifying messages of secular Indian identity and unwittingly contributed to the communalization of tensions in the 1980s. However, the success of Al-Jazeera in the face of competing Arabic satellite news services has elevated the stature of Qatar and increased its leverage on the global stage. In all these cases, in an environment of new communications options, government tried to take advantage of them to expand their information power network. In Canada and Qatar, the government increased its influence; in Brazil, the government failed to extend its reach; in India, the governments succeeded in reaching a broad audience but inadvertently fueled a divisive political trend that challenged the orthodox national identity.

Chapter 5 expands on the trust community concept and the implications for research. The epilogue applies theory to practice.

Communications transform politics when a series of changes occur. Technology helps connect more people together who would otherwise not be connected. Then, when more people are connected, there is a chance for them to improve their understanding of each other and to build common norms and values. Finally, this trust helps build social capital, one of the essential resources of a group that enables them to act together in the interest of the community.

This book connects two worlds—the sphere of political action, and the sphere of information and ideas. It will be apparent throughout that those with the information and ideas can influence the world if they use appropriate communications tools to distribute them. Change in communications technology and massive flows of new information result in significant political transformation if the people who use the technology re-imagine their own identities and if new communities emerge in society to compete with the old ones. With access to new information and different views of the world, people can see themselves in an altered light. In some cases, knowledge of lifestyles in other parts of the world affects the local lifestyle. In other cases, these new views of the outside world are not accepted as good, and an opposite reaction occurs, an ever-tightening adherence to local values and customs. If we accept that information is a source of power, then understanding how it is used, distributed, and interpreted is an essential aspect of understanding politics in a given community.

CHAPTER 2

Blogs, Wikis, and International Collective Action

The 2004 Boxing Day Tsunami

CAN PEOPLE WHO only know each other online trust each other enough to work together? The *TsunamiHelp* case is an example of trust and collective action in a purely online environment. Revealed through old-fashioned fieldwork, the story of the start, growth, and fading away of this online community shows that it is not so very different from that of unmediated communities. Equally, the recipe for successful collective action is not simpler online than offline, and the obstacles that challenge collective action in general also apply when new technologies are involved.

On December 26, 2004, the two largest earthquakes of the previous forty years ruptured a fault extending from Myanmar in the north to the islands of Indonesia in the south and westward to India, Sri Lanka, and the coast of Africa. The tsunami triggered by the quakes struck eleven countries in South and Southeast Asia, killing more than 225,000 people.[1] According to the U.S. National Oceanographic and Atmospheric Agency, this 2004 tsunami was the most deadly since the Calcutta, India, earthquake in 1737 (see fig. 2.1).

In India a small group of bloggers responded by creating a blog (http://tsunamihelp.blogspot.com) that became a wiki (www.tsunamihelp.info/wiki/), which then became a global clearinghouse for people who wanted to help tsunami victims. People posted what they knew, what they had, who needed what, and what was needed where. Within its first week, the *TsunamiHelp* blog was a top humanitarian website, just behind the United Nations, Reuters AlertNet, and the

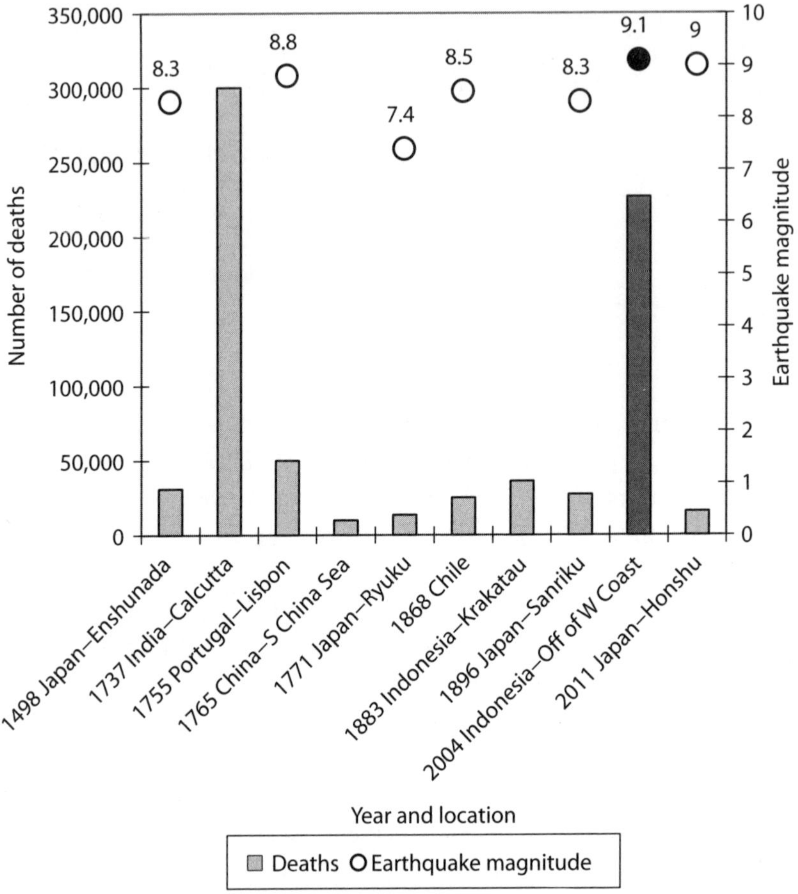

FIGURE 2.1. Most severe tsunamis by mortality, 2000 BC–2012 AD.
Source: National Geophysics Database, NOAA.

International Federation of Red Cross and Red Crescent. In the United States and the UK, *TsunamiHelp* was the tenth most-visited humanitarian site that week; in Australia, it was the fifth.[2] In contrast, government and other institutions failed to respond quickly and were inept at using technology to crowdsource information. Ordinary citizens used online technology to express anger, frustration, and grief.

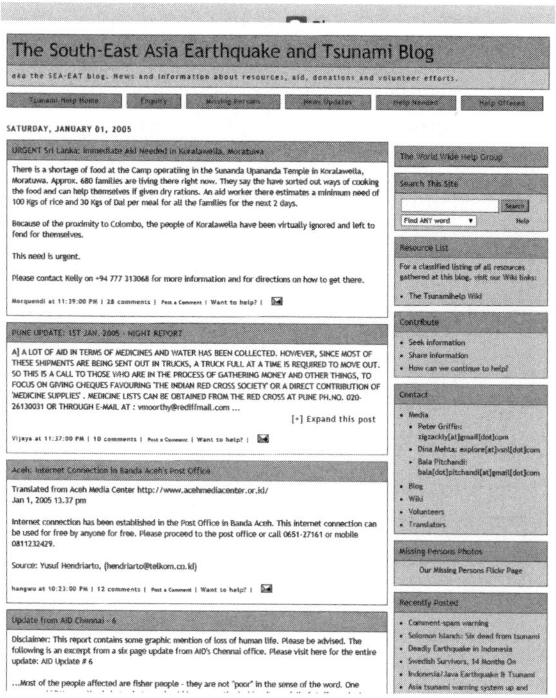

FIGURE 2.2. *TsunamiHelp*: blog screenshot.

On the *TsunamiHelp* blog, people on the ground reported what was happening, who needed help, and how best to help in all the affected areas. Word about the blog spread through the volunteers' personal networks, reporting by international media, and support from Internet companies. It grew not only because people read it but also because people wrote for it (see fig. 2.2). Eventually, the blog became too long—it was common to have hundreds of postings a day—and the information was organized by topic on a wiki. Volunteers both kept the blog running and maintained the wiki. The height of the *TsunamiHelp* campaign was the two to three weeks immediately following the disaster, when there was scarce information from other sources and worry about the victims was the most urgent.

Subsequently, several leaders of the group created similar blog and wiki campaigns focused on other disasters and causes.

The volunteers of the *TsunamiHelp* blog and wiki came together in anger, disappointment, and frustration with government institutions' inadequate response to the needs of the disaster victims. This became their common cause and shaped their common identity. The volunteers connected with each other online through emails, blogs, and wikis but also used newspapers, radio, and television to publicize their efforts.

Doing work together opened opportunities for volunteers to reciprocate good will, build trust, and create social capital. The main work was collecting information on disaster conditions, victim needs, and offers to assist as well as connecting victims with family and friends. The blog and wiki required cooperation among volunteers with diverse backgrounds, cultures, skills, and geographic locations.

The cache of information stored on the blog and wiki was the basis of the group's political power. With this information, they attracted millions of visitors. Once they demonstrated that their sites were central hubs of information, some international humanitarian groups cooperated. Within three weeks of the tsunami, the volunteers created one of the most visited websites of its time. By this measure, it was an effective collective action.

While *TsunamiHelp* volunteers successfully developed a trust community, it is harder to measure the effectiveness of their humanitarian efforts on the ground. The blog and wiki effort had a lasting effect on several volunteers. Many who had never volunteered before became regular volunteers in similar subsequent online efforts. Although individuals developed lasting skills, whether the group could have become an institution with formal rules and organization remains unclear.

How This Case Came to Light

For this research, I interviewed volunteers who worked on the *TsunamiHelp* blog and wiki. Several volunteers sent me batches of the email correspondence among the volunteers that document the coordination needed and challenges involved in constructing both the blog and wiki. I also studied the content of the blog and wiki, both of which were still available on the Internet at the time I began this research in 2008. Also, two volunteers wrote about their experiences very shortly after the crisis, Paola Di Maio in 2005 and Peter Griffin in 2007.[3]

In 2004, at the time of the tsunami, blogging was already established as common Internet activity; building a wiki was something relatively new. Blogging was the online equivalent of keeping a diary, ideal for sharing personal stories. By 2004, blogging made it easy for authors to publish content and for others to comment. Bloggers could use the free services offered by Internet companies, which were in turn supported by advertising. The structure of a blog was bound by its history as an online diary. On the computer screen, blog posts show in reverse chronological order. Every author's post remains a distinct unit, and authors generally cannot change their chronological organization. The author controls the content of a blog post or a comment; no one else can edit it, in contrast with a wiki.

The wiki is more flexible than the blog. A wiki's simplified commands allow users easily to edit the content and the organization of the page. Administrative control of wiki pages can be assigned to an individual, a group, or, in the case of the *TsunamiHelp* wiki, to the public. Such flexibility means wiki pages are easier to organize clearly than a blog, but the authorship or origin of the content is less easy to trace. Users can create wikis from free services offered by Internet companies.

On the *TsunamiHelp* wiki a contact list was posted for visitors who wanted to know how they could help. About two dozen names were listed. I contacted all and was able to interview six in person and ten by telephone between January 2008 and September 2009. After most of the interviews were complete, I compiled the information the volunteers had shared with me and posted them to a blog called *volunteerpoweronline* in August 2009 (www.volunteerpower online.blogspot.com). Within days of posting the information, several volunteers returned to me with corrections and edits to the information. Also, one of the blog's original creators, Peter Griffin, sent news of the blog's creation to a network of bloggers. A few additional people volunteered to be interviewed through this process. A full list of interviews is in the appendix to this chapter.

Failure of Government, Humanitarian, and Media Institutions

> Tsunamihelp.blogspot.com. Saturday, January 01, 2005. URGENT Sri Lanka: Immediate Aid Needed in Koralawella, Moratuwa. There is a shortage of food at the Camp operating in the Sunanda Upananda Temple in Koralawella, Moratuwa. Approx. 680 families are living there right now. They say they have sorted out ways of cooking the food and can help themselves if given dry rations. An aid worker there estimates a minimum need of 100 Kgs of rice and 30 Kgs of Dal per meal for all the families for the next 2 days. . . . Because of the proximity to Colombo, the people of Koralawella have been virtually ignored and left to fend for themselves.

The *TsunamiHelp* moment was in the immediate aftermath of the disaster when news that it had occurred had spread but there was little on-the-ground media, government reporting, or international humanitarian effort. Several volunteers remember searching for information about the tsunami and finding nothing. Megha Murthy in Boston and Nancy Bohrer in Chicago both turned to the Internet because they could not find good reporting from traditional media.

In the first days, Peter Griffin, Dina Mehta, and Rohit Gupta in Mumbai were at the center of the blog. The official name of the email group that started the blog was South-East Asia Earthquake and Tsunami (SEA-EAT) blog, at tsunamihelp.blogspot.com. Prior to the tsunami, Peter and Rohit were collaborating on a literary blogging effort—*desimediabitch.blogspot.com.* They were just launching the site when the tsunami hit. Peter and Rohit turned their literary community toward disaster relief. Dina Mehta, an early, prominent blogger in India, reached into the blogging community to establish *TsunamiHelp*.

At the time, blogging was the only online social network application available that allowed users to contribute content. Also, since blogspot.com had just been acquired by Google, *TsunamiHelp* volunteers had access to a free system of resources with enough scale to sustain the flood of hits to the website.

Word spread about the *TsunamiHelp* blog. In Mumbai, Peter Griffin, Dina Mehta, and Rohit Gupta gave radio interviews. The London newspaper the *Guardian* picked up the story. Bala Pitchandi in New Jersey and others said that Google featured the *TsunamiHelp* blog on its front page. All this coverage triggered a flood of volunteers. Suhit Anantula in Hyderabad, India, was between jobs and waiting to leave for Australia. He knew Dina from some time he had spent in Mumbai. When Dina asked, he joined. Andy Carvin in Washington, D.C., knew of Peter and Dina through *globalvoices.org*, an American web-blogging effort. Katherine Bertolucci, at her parents' house for Christmas in the United States, received a message asking for help on the *TsunamiHelp* blog from Paola Di Maio through an email list for content management professionals. The most intense activity was from the end of December 2004 until mid-January 2005. After that point, traditional media outlets and traditional humanitarian institutions were in position to report.

Why Individuals Came Together and How They Did It

> Tsunamihelp.blogspot.com. December 31, 2010. Find Missing Persons in Myanmar (Burma). The Center for Diplomatic Missions has staff on the ground in Myanmar (Burma) and is willing to relay missing persons inquiries for local effort to locate by phone or address visit. Send the best available data regarding the person sought, include all hotel names, cities and any other data you may have or suspicion—the more

> information you provide the better our staff can assist you. Communications in and out are difficult-to-impossible, but we have Satphone contact with email relay and will be happy to assist as many as possible. James A. Howell, Director, The Center for Diplomatic Missions

All the volunteers were shocked at the magnitude of the disaster, angry at government and nongovernment institutions' failure to warn and rescue people, and frustrated at the lack of good information and effective outlets to help the aid effort. They felt grief for the victims, an impulse to contribute beyond money, and exhilaration in working together for a cause.

SHOCK, GRIEF, AND ANGER

"I was so helpless," said Bala Pitchandi, "When most of these disasters occur, people feel helpless sitting wherever they are. While economically you can contribute . . . if you do something yourself that would help find that someone is OK, that would be better than sending a check." Bala, a software engineer in New Jersey had family in Tamil Nadu, India. For a short time, he did not know whether his parents, whom he knew were visiting a beachside temple that day, might have been hurt. Fortunately, they were safe, but he knew others who were deeply affected. Online he came across the *TsunamiHelp* blog and sent Peter an email. In the beginning, Bala's knowledge of people and local news sources in the disaster area was immediately helpful. Later on, he played an important role in coordinating work among the volunteers and continuing on in subsequent humanitarian campaigns.

University lecturer in information systems Paola Di Maio lived in Thailand just a few kilometers from a beach hit by the tsunami. The earthquake hit at 8:00 a.m., and a half hour later a large wave hit the beach. Without television or phone, she relied on email to alert her parents in Italy that she was safe. Then she saw from a *RedHerring.com* news feed that an online group was forming. A mailing list started about twelve hours after the quake. With her experience in system engineering, she helped structure the blog.

Student Angelo Embuldineya, son of a Tamil father and Singhalese mother, had just returned to Bahrain to resume university when the tsunami hit his native Sri Lanka. Previously, while in high school, he had volunteered with the United Nations Development Program (UNDP). When the tsunami hit, he returned to Sri Lanka. The UNDP contacted him, and he started working to help the victims. Through that work, he came into contact with Bala Pitchandi and Dina Mehta, who introduced him to *TsunamiHelp* group in January. While his own family was not hurt, he had friends who were affected. He knew two people who died because help had not reached them quickly enough. He knew many people who were injured. The level of aid available and the red tape involved in providing it, he said, was appalling.

FRUSTRATION AND THE IMPULSE TO HELP

Most volunteers helped, even though they were not directly affected themselves. The blog and wiki expressed their grief and desire to alleviate suffering, a statement about their common humanity. Megha Murthy in Boston said, "To financially help only takes a few minutes. . . . What I could contribute was time—time is not money. That was really the only driving force; I wanted to help, helping with money was just not adequate." Megha is originally from India; when she heard about the disaster, she used Google to search for "tsunami," and the *TsunamiHelp* blog came up as a result. When she received Rohit Gupta's message to the email group asking for help with the blog template, she volunteered. Megha is a website designer and brought professional technical knowledge to the work. Katherine Bertolucci in Phoenix, Arizona in the U.S. said, "You hear about these horrible disasters and want to help out, but can't. I'm not a nurse. So here's an opportunity, so I said yeah I would help. I told her [Paola] that what I did was organize information. It was exciting, it was needed, and I was really participating in helping people. It was just terrible, tens of thousands of people dying in different places and here we were sitting in comfortable offices. What can I give them? I can give them hours."

Susan Turnbull in Washington, D.C., was the first to tell me her story of volunteering for *TsunamiHelp* and shared the trove of emails that documented its management. She and Dina both believe that the humanitarian impulses were not fully tapped by governments and aid agencies. *TsunamiHelp* was a success because it required people to give, not take. It turned readers into contributors, spectators into agents. In Amsterdam, Rudi Cilibrasi echoed this sentiment. He saw the wiki as a way for people to leave their mark, to express their grief and their anger. Volunteers were exhilarated by joining hands across the globe. Anna Lissa Cruz thought it was cool that they were able to create these blogs as a team—she and Rudi in Amsterdam, a bunch of people in India, Bahrain, and the United States. It was just incredible, she said, that they were able to do this even though they were thousands of miles apart.

Creating the Blog and Wiki: Building a Repertoire

> Tsunamihelp.blogspot.com. January 6, 2010. Finding Information in the Blog Search. There have been tons of information being posted on this blog so if you are looking for specific information, the easiest way to find information is to use the search facility on top of the side bar of this page. You can search this blog, all of our sister blogs (accessible through the buttons above) and our Wiki pages as well.
>
> Wiki. To help find information better, we have created a classified Resource List we call Wiki Tsunami Help Portal. They are better organized and categorized into Aid Agencies, Helpline Numbers, Fundraising Events, Missing & Found, Health & Safety and Ground Zero Information etc.
>
> Comments/Suggestions/Questions. Your Suggestions/Links go here. Your questions/info to the bloggers go here.

Disclaimer. We do not endorse any of the listed organizations either here on the blog or on the comments left by our readers. We are simply here to help you find the relevant information.
—Bala

Sidney Tarrow observes that in France between 1780 and 1850, protests evolved from unorganized riots to professional protests involving the construction of complex barricades across streets and thoroughfares. Repeated practice had inscribed on the community's collective mind the knowledge of how and where to build the most effective roadblocks. Such techniques could be applied at any time for a protest of any sort. In short, building the barricade had entered the repertoire of Parisian's collective action techniques.[4] For the participants in *TsunamiHelp,* working on the blog and wiki was like learning to build the barricade. Much of the volunteer work was quite tedious—the functional equivalent of stuffing envelopes, constructing signs, and distributing flyers. Bala said his role was akin to traffic police. Volunteers would contact him; he would welcome them and assign work on the blog or the wiki. While some had volunteered before, for several this was their first online volunteer effort. Many of the struggles they discussed with me were about the problems of building barricades—in this case, making the blog easier to use, organizing the wiki, and, finally, finding a new home for the wiki.

MAKING THE BLOG EASIER TO USE

In the beginning, there were too many blog posts. The default blog format posted the most recent contribution at the top of the screen, moving all previous contributions down the screen and eventually onto a "next" less-visible page. Very recent postings did not stay on the front page very long and were quickly buried in second and third pages. Megha saw that the blog was hard to navigate. The volunteers formed a group to redesign the blog but failed to reach consensus. In the end, Megha proposed a blog re-design and asked for comments. Rohit, Dina, and Peter were pleased, and Megha implemented the re-design on December 30, just a few days after the blog had started. Megha's major innovation was that individual posts could be collapsed, allowing a reader to view more posts on a single page view than before. About a hundred posts could then be viewed at once on the main page. This was an important innovation because many readers of the blog only had access to a slow dial-up Internet service.

CREATING THE WIKI

As the blog grew, readers had trouble finding the information they needed. The volunteers experimented with creating sub-blogs, but this was clumsy. So they turned to wikis, still a very new concept. Suhit Anantula in Hyderabad told me that he started the wiki. To convert the blog into the wiki, the information had to be categorized. For the first two or three days of the wiki, volunteers focused on bringing order to the information. Taxonomy expert Katherine Bertolucci organized information in alphabetical order or by country. Someone would email information to her from the blog, and she would be given a wiki page to organize. Rob Kline, a software programmer who worked on the *TsunamiHelp* wiki anonymously, estimates he analyzed about three hundred blog entries, arranged them topically, and then helped maintain the wiki by continuing to check if the blog had interesting posts to transfer. Like Katherine, he organized information along two paths—those who needed, and those who gave. He also clustered information by interest—geographically or by organizational interest, for example, to purify water (see fig. 2.3).

Bala remembered a volunteer from the BBC who made a mobile version of the blog available. Anna Lissa worked on a database with lists of people who were thought to be in affected areas. It allowed others to enter information about their status: missing, found, or deceased. Another database listed resources and how they were being used. Nancy worked on the wiki's Ground Zero page, collecting in-

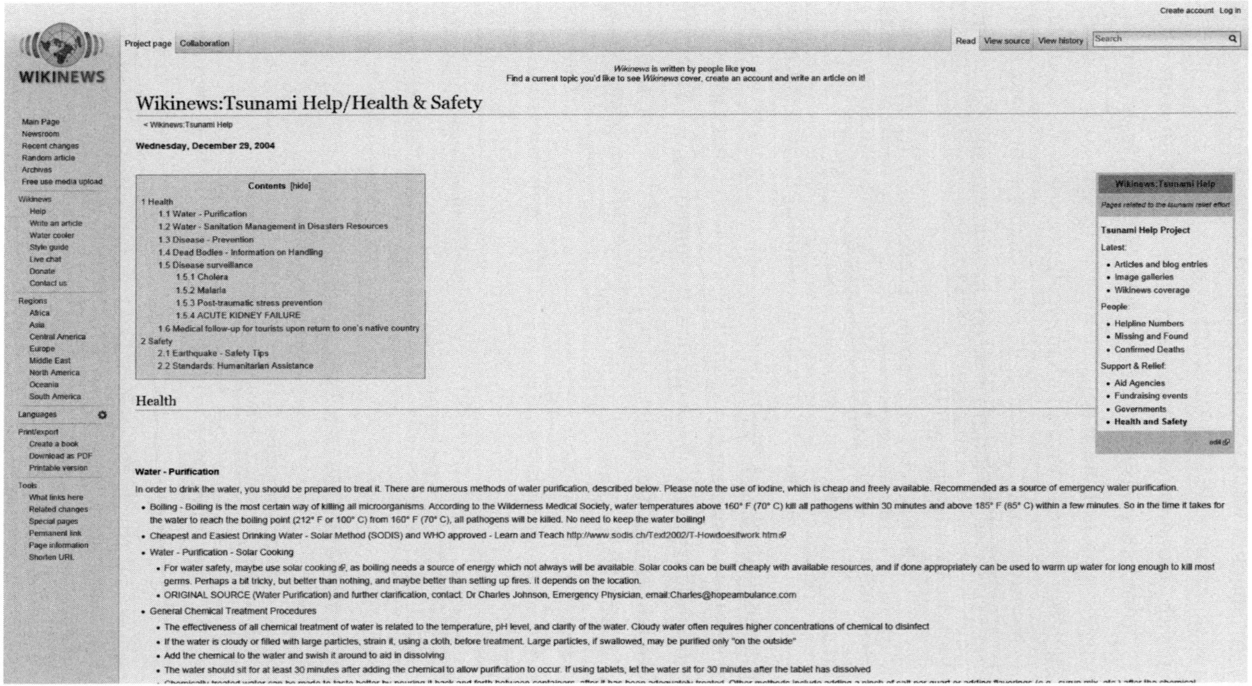

FIGURE 2.3. *TsunamiHelp:* wiki screenshot.

formation from other websites on the relief effort to enable others to find out quickly what was happening. Angelo, who was in Sri Lanka at the time, sent SMS (short messaging service) to a phone number with "tsunami info" as a prefix, with information on what help was needed; this was matched with help offered. Megha remembers spending a lot of time debugging information as posts went up on to the blog. As Neha Viswanathan recollects, volunteers took on roles as "monitors" and "janitors," who watched for duplicate posts on the blog, classified the information on the blog into sub-blogs, and cleaned up the information on the wiki.

"It was a 24-hour-a-day operation. There was a feeling of urgency—if you went to sleep, you were not helping save someone," said Katherine. Likewise, Suhit remembers sleeping only three to four hours a day for about four to five days working on the wiki. Suhit used chat rooms to get the wiki going and tasked out work to small groups of people. Sitting in Hyderabad, he could feel that there were people around the world who wanted to do something, who wanted to contribute, who just needed to be channeled to the right area. He recalls working with volunteers from Australia and with people from all time zones. Katherine recalls working with a lot of college students in America who hailed from affected countries and wanted to volunteer. While those interviewed could not pinpoint exactly how many volunteers participated, the wiki drew together dozens, if not hundreds, of volunteers.

MOVING THE WIKI TO A NEW HOME

"We needed a wiki; the most obvious [homes] were Wikipedia and Wikinews," said Bala, recollecting the practicalities of getting the wiki established. "The rules of Wikipedia were that only facts could be posted, only neutral information—[it was] not all right to post information that this is the hospital to call for missing person information. Therefore, we thought Wikinews would be the appropriate place. None of us had installed a wiki on our own before. The *TsunamiHelp* wiki generated so much traffic that it initially brought down Wikinews." Then, slowing down development further, Wikinews's editorial policy of maintaining a neutral point of view on the news meant that much of the humanitarian-oriented information on the *TsunamiHelp* wiki was not considered appropriate by the Wikinews administrators.

It fell to Bala and Dina to talk to the administrators of Wikinews, and they decided to move the wiki to a new server. Wikinews helped the volunteers move the wiki. Dina bought the URL and server space from GoDaddy, and eventually this was handed over to volunteers Rudi Cilibrasi and Anna Lissa Cruz, a couple in Amsterdam who had the server capacity to host the wiki. The wiki continued in their care until mid-2009.

From a technical standpoint, Rudi's main concern was managing the level of traffic to the wiki site. He said he had had one million hits the first day of the wiki, and there were several days like that. Once, the server overloaded, crashed, and had to be rebooted—taking the wiki out of commission for several hours. Spam was another problem. Finally, there was the cost of bandwidth. By Rudi's accounting, the server that hosted the wiki cost US$10,000, with a monthly charge of US$150 per month for Internet service bandwidth. Through the wiki, Rudi remembers collecting about US$2,000. The same server supported some other projects of his, and as of late 2009, he had collected an additional $1,000 from other philanthropic and not-for-profit sites he has supported.

Creating the wiki content was the accomplishment of many volunteers who largely did not know each other personally and were not organized in any particular hierarchy. No one I interviewed clearly identified any single person as responsible for the wiki. Volunteers specialized in self-selected areas where they thought they could be useful. The first challenge, the compression of the blog postings, was essentially the work of Megha in collaboration with the group. There is little question that she was responsible for the technical solution. The final challenge, the hosting of the wiki, was a problem solved by several volunteers who are acknowledged as leaders by others. These leaders were scattered across the time zones—Bala in the United

States; Rudi Cilibrasi and Anna Lissa Cruz in Netherlands; Dina, Rohit, and Peter in India. Except for Rudi and Anna Lissa, who were engaged to be married, and Peter and Rohit, who had worked together in India, people in this small group had not worked together previously. Nevertheless, they were able to generate enough trust to accomplish the move from Wikinews to Rudi's server.

The Egalitarian Ethos: Philosophical Underpinnings of the Group

> Tsunamihelp.blogspot.com. January 12, 2010. It's funny, y'know. With so much corruption and everyday occurrences that reflect a lack of compassion from people, it is easy to lose faith in the human race. But what has happened to renew my faith has been the immense amount of love and support from close friends *and* acquaintances as well as the community at large. It seems that everyone has donated something to the cause, whether it be money, supplies, or food—and if they didn't have much to give, the amount of volunteering has been heartwarming. The aftermath of the tsunami has made many people's lives hectic with organizing aide to those countries affected, and unfortunately there hasn't been time for much else. I'm writing this letter to say thank you to all of you who have shown their concern, empathized with and prayed for those of us who are going through this process of mourning—your love is not ignored and will not be forgotten.—D'Lo

Many studies on transnational social movements feature organizations with a small professional leadership who use information technology to mobilize an extensive passive mass of supporters to participate in the occasional mass demonstration. These are organizations such as Greenpeace, which claims millions of members who largely contribute funds, but only a handful of professional activists carry out protests.[5] In contrast, the *TsunamiHelp* group is smaller in scale than that of Greenpeace, and the leadership is entirely volunteer. Furthermore, the group ethos was that everyone was equal and decisions would be made by consensus. Recent studies on hackers confirm that this nonhierarchical, consensus-driven approach is common in online communities.[6]

When Megha designed a template for the blog, there was no formal process for making a final decision. She had expertise and put herself forward as a leader; she deferred to others in areas where she did not have expertise. Katherine, who only worked on the wiki, said that although volunteers were assisting her, she did not consider herself the boss but only the conduit between the people who were working and the work that needed to be done. More skeptically, Nancy said that the group structure was both its genius and its failure—if you needed help with your work, it was your own job to find it. Compared to her volunteer work off-line, Nancy said the *TsunamiHelp* volunteers were good at running themselves and did not particularly like running others.

Within this egalitarian community, volunteers established credibility over time, said Megha. While initially it was challenging to understand who the others were and it was awkward to question people's credentials without a good reason, "By not knowing anyone, you had to trust everyone equally. Our unity was only in the cause," said Megha. Once a volunteer started to deliver work, however, then the work spoke for that volunteer. Others would listen to her, Megha said, because they recognized her capabilities. Katherine echoed Megha's sentiment: "In a volunteer situation, once you prove your stuff or that you're going to show up every day, that's the nature of it. If you work and do good work, you tend to get more responsibility." To prove her usefulness as a volunteer, Katherine purposely took on complicated problems and solved them first. Then people were willing to work with her to organize more material. Dina brought another angle to this process of building credibility. She maintains that when one blogs a lot, over time it is not possible to fake entirely one's personality.

However, not everyone was positive about the egalitarian nature of the group. Rudi, who had to maintain the wiki server, felt that work only was done if individual volunteers got inspired. He attributes the lack of work ethic to, as he put it, "a strange cultural norm . . . that there was a commitment to having no leaders." For the wiki, Rudi recalled there were 40 to 50 people involved in keeping different sections clean because it was under constant attack from spammers. Initially, volunteering worked well, but as initial shock of the disaster waned, there was more drudgery and it was harder to maintain the wiki.

Interestingly, while nearly all the interviewees asserted the egalitarian nature of the blog and wiki work, there was still widespread agreement that Dina, Peter, and Rohit had started the blog and that Bala Pitchandi was very important in carrying work forward. Several mentioned Megha as responsible for the blog template and designs that make the effort effective. It was also clear that Paola initially had a leadership position but was gradually excluded from the group. Paola, who lives just a short distance from the beaches in Phuket, was the only person in this circle physically located near an area hit by the tsunami. Some people worked well with her and others did not. Interviewed four to five years after the event, many volunteers acknowledged that conflict may have occurred in working with Paola because volunteers' nerves were frayed from lack of sleep and anxiety in the early days of the blog and wiki.

Blog and Wiki Effectiveness on the Ground

Rudi Cilibrasi exclaimed, "Can we give the name of one person whose life we saved? The answer was no, and that was sad." While recognizing the natural limits of a volunteer group, Rudi argued that the lack of process and a clear leader meant that it was hard to deliver aid when requests for help came in. This egalitarian structure left an accountability gap. Rohit Gupta, one of the blog's initiators, also doubted the blog and wiki actually accomplished anything. He was angry at the incredible failure of democratic governments to warn people of disasters. For him the blog and wiki were an alternative bureaucracy substituting for failed government and media.

Nancy said that not understanding who actually was helped was a true problem for *TsunamiHelp*. Familiar with other charities, she was conscious that the online effort never had a clear methodology for collecting information on actual effectiveness. Anna Lissa noted that while the volunteers were very passionate, smart, and opinionated, they did not necessarily have the tools and knowledge to do work. The technical challenges were left to Rudi and her. On balance, she was still happy to be involved in helping the larger goal of aiding people.

Rob was also concerned about effectiveness. Even as he went on to orchestrate a similar online effort for Hurricane Katrina just a few months later in 2005, one of his main frustrations was that it was difficult to learn more about the outcomes of their efforts. He was reduced to unique hits as a metric for the site, acknowledging that it is difficult to know whether readers found the information useful. For Rob, however, ultimately the effort was still worthwhile.

Neha remembered having an online chat in which in one window was a company that wanted to help and in another were people who wanted boats because they had lost their own and fishing was their livelihood. She matched people who wanted to give with people who had a need. Even with this concrete experience, she expressed some doubt. Were the boats ever bought and delivered? Yes, she thought, but it was a matter of trust.

Encountering similar challenges but with a more positive outlook, Bala recalled one instance when he heard from a group of doctors who were willing to travel. One of those doctors posted a message to the blog. With that message, volunteers contacted a local agency in Sri Lanka and people in relief agencies in Thailand in the area where Paola was located. The volunteers helped the doctors get in touch with relief agencies. Morquendi, a blogging friend of Rohit Gupta's, sent concrete information through *TsunamiHelp* to aid agencies. Morquendi did not have regular Internet access after the tsunami but was able to send frequent SMS messages by cell phone, which Rohit posted to the blog. Morquendi's reports were about the destruction, injury, and deaths he saw on the beach. The government was slow to report, and the traditional news media had yet to reach the area. Angelo, who flew back to Sri Lanka after the disaster, also sent SMS information on missing people. He recalled using information from the blog to help Rotary International get the needed U.S. military air support for tents and food drops when the Sri Lankan military was overextended. For the work in Sri Lanka, United Nations relief effort gave Bala access to their own communications because they found the blog and wiki's on-the-ground information valuable.

"I felt if there was just one family helped, that was enough. There wasn't the certainty of helping someone. . . . I just hoped that I had done good," said Megha. Volunteers like Dina and Megha said they felt a greater sense of control over their contribution when it was in-kind service to the blog and wiki. Money contributions, they felt, often ended up in the wrong hands. Their time on *TsunamiHelp* was relatively more certain to help than financial contributions alone.

In short, the practical effectiveness of the *TsunamiHelp* blog and wiki is unmeasured and therefore unclear. Nevertheless, most volunteers interviewed were enthusiastically certain their efforts benefited tsunami victims. The blog and wiki may be best measured, not in comparison to efforts by other charities, but as an expression of frustration with the traditional media, political, and humanitarian organizations—a protest against the media and government status quo. For some, time put into the blog and wiki was a substitute or complement to financial contributions when they had doubts as to the true purpose of the fund recipients. For others, time put into the blog was a shout of anger against the shortcomings of government authorities in preparing for disaster. For yet others, information poured into *TsunamiHelp* efforts filled a vacuum left by a media completely unprepared to cover remote parts of the world.

Did Volunteers Participate Again in Other Collective Actions?

Were the identities of the *TsunamiHelp* volunteers transformed by the experience? Did this online activism build an organization that would act again? Studies of transnational movements often address whether the personal relationships between participants engender trust, create a collective identity, and communicate needs enough to grow from a network to a social movement.[7] Are the participants' interests and identities redefined enough to lift a network into a movement?[8]

In the case of *TsunamiHelp,* most of the volunteers interviewed identified it as a life-changing experience. Only two of the sixteen were disillusioned with the movement. Those interviewed were the ones most deeply engaged in the work. There are dozens, perhaps hundreds more volunteers whose views remain unknown. For several of the interviewees, the experience was so transformative that they went on to participate in other collective actions—humanitarian help for other disasters, protests against Indian government restrictions on blogging, or creating artistic or literary communities online. For a few volunteers, *TsunamiHelp* represented a new technique in their repertoire of volunteer and activist work. For others, *TsunamiHelp* was an initiation into being part of a social movement. Finally, for two, Rudi Cilibrasi and Rohit Gupta, the frustrations of *TsunamiHelp* work were so great that they doubted the benefit of the exercise in light of the effort expended.

Several of the interviewees had previous volunteer experiences. Katherine had volunteered in collecting goods for relief efforts. The kind of work she performed for the wiki—the systematic classification of information—paralleled the kind of volunteer work she did in the off-line world, such as the systematic classification of donated goods. Anna Lissa had volunteered before at a science museum when she lived in San Francisco and had participated in rallies. Anna Lissa said in her view that as long as one person was helped by the effort, the effort is worth it, a statement that encapsulates the motivation of people who regularly volunteer. For many years Nancy has worked on a local charity sale. With her experience in this highly organized volunteer effort, she acknowledged that the *TsunamiHelp* group lacked the kind of systematic accounting of resources and results that are the foundation for charities' reputation for effectiveness.

However, from her interaction with users of the sites at the time of the crisis, she personally was certain the blog and wiki did help people. Volunteers like Katherine, Anna Lissa, and Nancy brought to *TsunamiHelp* a well-developed and well-practiced ethos of volunteering for the community.

However, for many of the interviewees, this was their first volunteer experience, and it set them on a path to volunteering for other causes as well. Dina said the *TsunamiHelp* experience strengthened her belief in the media and in the opportunity for a person to make a difference. She organized an online protest in 2006 against the Indian government's ban on blogs. Neha joined Dina on this campaign. Bala subsequently worked on *quakehelp.info* after a major earthquake hit Pakistan in October 2008. Megha started working with Peter on other online projects, such as the website for the Kala Ghoba art festival every February in Mumbai. Suhit Anantula claimed he learned more in the ten days of working on the wiki than he could have learned in a year. Never before, he said, had he experienced the power of collaboration. When interviewed four years after the experience, he was still seeking to replicate it in other areas of his life and work.

For others, the *TsunamiHelp* experience also transformed their professional lives. For Neha, the *TsunamiHelp* experience was a turning point. Previously a social worker, she started working as regional editor for South Asia for the Berkman Center's Global Voices. She became professionally interested in blogging. She began to see how blogs could not only increase transparency and make ideas more effective but also exclude people in the process. Bala also said that the *TsunamiHelp* work changed his life. The tsunami experience underscored for him that there were a lot of good people all around the world who wanted to help each other. He started participating in *newcomfarm.com,* a conference that looks at the implications of new communications tools. It has given him a different way to be engaged with the world, to be involved in international development.

In August 2005, just eight months after the tsunami in Asia, Hurricane Katrina devastated New Orleans in the United States, a catalyst for the most consequential post-tsunami collaboration among the volunteers. Rob Kline, who was vital in instigating this work, said he and others essentially cloned *TsunamiHelp* for Katrina as they were awaiting its landfall. He sent the idea out to the *TsunamiHelp* team, and within three hours they were up and running. The major leaders came from the *TsunamiHelp* group—Bala, Nancy, and others. Rudi in the Netherlands started the site on his server in Amsterdam; later they moved it to the United States. Dina, Anna Lissa, and Megha all participated. *Katrinahelp* was up and running before the websites of the Red Cross and the U.S. Federal Emergency Management Agency (FEMA). For the Hurricane Katrina work, the *TsunamiHelp* group developed a model that included working with the Red Cross to develop a similar missing links project to help people find people.

By contrast, Rudi Cilibrasi's deep technical engagement with *TsunamiHelp* left him frustrated at the modesty of its achievements. While it absorbed a huge amount of time and resources from the volunteers, the *TsunamiHelp* work, he felt, could not really point to any concrete benefit or result for victims of the tragedy. When he discussed the experience five years after the tsunami, he was skeptical that such volunteer work had much actual impact and seemed unlikely to be involved in like work again. Despite his protestations, some time after interviewing him, I found that he had participated in an online effort to help humanitarian efforts after a major earthquake hit Haiti in early 2010.

For the individuals who engaged, it left a deep mark, although the entire episode of the *TsunamiHelp* blog and wiki was brief and ephemeral. For most volunteers, the effect was a positive one. Volunteers with past experience had their values of giving and participation reinforced. Many who were new to volunteering carried this experience forward to more volunteer efforts online. A minority was disenchanted with the lack of empirical evidence that their online efforts had any real benefit to victims of the disaster.

As an institution, the *TsunamiHelp* group had at least one clear reiteration in the effort around Hurricane Katrina. Nearly the same leadership infrastructure and the learned skills of the wiki and blog were replicated several months later for a disaster that hit a completely different part of the world with quite different resource constraints and needs. This suggests that *TsunamiHelp* was critical to the development of a repertoire of techniques for online activism. On balance, however, the greater effect was the transformation of individuals and their sense of empowerment rather than the building of a potentially new political institution.

The View through the Lens of a Trust Community

The trust community of *TsunamiHelp* volunteers shows how strangers can come together, develop a common identity, establish trust through cooperative work, and build social capital. The volunteers progressed from being merely a network connected by technology to creating a trust community; whether or not they progress to building an institution remains to be seen. This suggests that in today's Internet world, some mediated communications can be as rich and politically significant as face-to-face communications.

CREATING A COMMON IDENTITY

Among the *TsunamiHelp* volunteers, the common trauma was the shock at the disaster, anger with the ineffectiveness of institutions, and frustration with options to offer assistance. These feelings were acutely evident in the emails to manage the blog and wiki and in interviews with the volunteers. Sharing the feelings of distress online created a basis for action.[9] As people expressed emotions and responded to each other, the volunteer network became a community with a sense of identity. From this consensus view, the volunteers sprang into action—in this case, collecting information on the disaster and humanitarian aid.

Scholars of collective action like Douglas McAdam and Sidney Tarrow speak of political opportunities that arise as catalysts for action. In the case of the *TsunamiHelp* blog and wiki, the new political opportunity that emerged was the failure of traditional media to provide adequate information on the disaster quickly enough. Increasingly, "citizen reporters," people on the ground, participants even, are able to provide information on breaking news events by posting to the Internet nearly instantaneously—inevitably more quickly than professional news reporters. In the case of a disaster like the large tsunami of 2004, patience for such professional reporting wore thin because the humanitarian needs were urgent and great and because many of the areas affected were particularly remote from global media centers. In addition, the applications available to the *TsunamiHelp* volunteers loosened constraints on collective action. Ordinary users—not just professional Web content producers—could easily create websites and post information to the Internet. In the wake of the tsunami, lack of information was the political opportunity. It was a moment in history when the costs and constraints of collecting and sharing information were evaporating but traditional institutions had not yet adapted.

BUILDING TRUST

People draw on their resources and repertoires to express opposition to a situation. In the case of the *TsunamiHelp* volunteers, the motivation to mobilize was to express grief, anger, and frustration—grief for the victims of the disaster, anger at the incompetence of governing institutions to provide relief, and frustration with the limited opportunities for ordinary people to offer assistance. Designing, discussing, building, and maintaining the blog and wiki were all activities that created opportunities for volunteers to reciprocate good will; they helped each other and learned to trust each other through this process. Through this work, volunteers also developed a repertoire of norms and routines to accomplish specific tasks. The *TsunamiHelp* campaign created a human network using old and new technologies to establish norms and work routines in the service of a larger goal.

To communicate with each other they used email, instant messaging, short message service (SMS) over cell phones, and Voice over Internet Protocol (VOIP). They also gave television, radio, and newspaper interviews to spread the word to the public.

One of the central questions in studying collective action is what kind of prior social network must exist, if any, for people to trust each other enough to cooperate? Doug McAdam and Florence Passy emphasize the importance of social networks that exist before the collective action, informing the action itself by providing blueprints for communication, hierarchies of leadership, and, most importantly, establishing lines of trust.[10] As McAdam put it, "Stable neighborhoods, churches, organizations are the best condition for collective action, not isolation, marginality."[11] If participants trust each other, they are more likely to make sacrifices as individuals in exchange for uncertain gains for the group.

The *TsunamiHelp* case is striking in that for the volunteers, when all these communications tools are used together, they can approximate the complexity of relationships built face-to-face. In the *TsunamiHelp* case, strangers were able to successfully conduct an international humanitarian campaign. Professional email lists, mass media reports, and Internet search engines connected the people in the *TsunamiHelp* human network. We should re-evaluate whether people must have prior relationships in order to build the trust needed for collective action. Remarkably, purely online interactions produced enough trust for collective action among people who were largely strangers. For a brief time, their campaign rivaled national government, international humanitarian organizations, and global media institutions. This success underscores the power of pure information, divorced from substantial economic or military resources. The implication is that online interactions have reached a level of complexity and richness that can foster the kind of trust previously attributed only to face-to-face interactions. Moreover, the *TsunamiHelp* case highlights the centrality of the technology-mediated human network—not institutions, the nation state, or even new technologies—in understanding collective action.

SOCIAL CAPITAL

To get something done, groups require social capital—a fund of goodwill and trust that enables people to cooperate. For the *TsunamiHelp* volunteers, once they trusted each other, they could act together to build the blog and wiki. They kept the blog clear of redundancies, organized information on the wiki, and monitored and janitored the wiki free of spam. They found ways to collect photos and information on missing people. This type of work is the kind of social knowledge, as Tarrow describes, possessed by the French activists who understood where, when, and how barricades are best built.[12] Enough trust developed to create the social capital necessary to act together not only for the 2004 tsunami but also for subsequent disasters.

NETWORK

While the blog and wiki were the technologies that enabled the volunteers to take advantage of the news vacuum, in order to work together they used all kinds of communications. Word about the group's work spread through newspaper reports, radio interviews, and professional email lists. In order to get work done, the volunteers used email, chat, and voice calls on Skype. The new technology amplified the message and introduced novel dynamics in the communications, but the other, older communications techniques still were important.[13] The importance of the blog and wiki is that they add to the available repertoire of communications tools. They should be considered not just as the latest Internet gizmo but rather as the latest way that people exchange information and ideas, a class of artifacts that extends to the television, the radio, the newspaper, the book, the parchment scroll, and the cave painting.

TRUST COMMUNITY

Initially, the network of volunteers was nonhierarchical. Volunteers came and went as they pleased; the group was held together by a common purpose, expressed in the collection of information. The volunteers trusted and cooperated with each other, and the community was able to adapt to outside pressures.[14] The *TsunamiHelp* volunteers all described the egalitarian nature of the community. Gradually, however, as the volunteers' activities grew in complexity, certain leaders emerged, even though at the time of the interviews, these leaders were reluctant to acknowledge their status.

This "anti-leader, anti-celebrity ethic" is a hallmark of Internet trust communities, as Gabriella Coleman documents in her initial sketch of the "hactivist" group Anonymous. She notes that within that community, participants perceived to be using the group's work for self-aggrandizement can be ostracized and exiled from future work.[15] Jennifer Brinkerhoff's work on Afghan and Somali diaspora communities online also confirms that voluntary communities are easy to join, easy to leave, and tend to be nonhierarchical and, in her cases, non-coercive.[16]

INSTITUTION

The *TsunamiHelp* episode at its most intense lasted for a couple of months. Afterward, a core number of volunteers continued to cooperate in similar responses, which suggests progress toward institution-building. However, as of 2010, the group did not appear to have established clear rules for work that characterize institutions. The strength of the *TsunamiHelp* group was its ability to cooperate to build a cache of information that was unrivaled by any other effort; the collection and dissemination of information was its main distinction, and its source of power. Whether this group can have a similar impact in other instances remains to be seen. However, even if this particular group does not again achieve the same impact as it did with the 2004 tsunami, individual volunteers have absorbed the ethos and techniques that can contribute to the emergence of new campaigns.

Appendix

APPENDIX 2.1 *TsunamiHelp* Interviewees

Family name	*Given name*	*Profession*	*Location in December 2004*	TsunamiHelp *work*
Anantula	Suhit	Business, blogging	Hyderabad, India	wiki
Bertolucci	Katherine	Library science, classification	Phoenix, US	wiki
Bohrer	Nancy	Attorney	Chicago, US	wiki
Carvin	Andy	Public radio product manager	Washington, DC, US	blog
Cilibrasi	Rudi	Internet and web programming	Amsterdam, Netherlands	wiki
Cruz	Anna Lissa	Web programming	Amsterdam, Netherlands	wiki
Di Maio	Paola	University lecturer, information systems	Phuket, Thailand	blog
Embuldeniya	Angelo	University student	Bahrain, Sri Lanka	blog
Griffin	Peter	Writer	Mumbai, Delhi, India	wiki and blog
Gupta	Rohit	Reporter	Mumbai, India	blog
Kline	Rob	Software programmer	Seattle, US	wiki
Mehta	Dina	Blogger	Mumbai, India	wiki and blog
Murthy	Megha	Website designer and developer	Boston, US	blog
Pitchandi	Bala	Software engineer	Hackensack, NJ, US	wiki and blog
Turnbull	Susan	Government technology expert	Washington, DC, US	wiki
Viswanathan	Neha	Social worker	London, UK	wiki

PART TWO

Network Technology Case Studies

CHAPTER 3

Activists Challenge Institutions with Information Technology Networks

IN 2011, EGYPTIANS and Tunisians used Twitter and Facebook to overthrow their governments. In the 1980s, Taiwan's political opposition used cable television to mobilize people to their cause. In the 1990s, nongovernmental organizations used phones, faxes, and email to run a campaign against landmines. In 2001, Filipinos used mobile phones to protest against a corrupt president and force his resignation. Going further back, in 1900, protestors in China used the telegram to organize against the Qing dynasty. Thus, for a long time, trust communities have used new technologies against the establishment to communicate ideas and information that convey a different perspective, an alternative vision. These visions are as powerful as geography, economic ties, or military force in holding a community together. Ideas and information are sources of power, and communications technologies are tools to exercise it.

This chapter explores how activists throughout history have used the newest communications technology to their advantage. I draw from scholarly works that usually have a different goal—demonstrating civil society in imperial China or exposing the history of corruption in the modern Philippines, for example. Instead, I mine these cases for very specific points. How did activists use communications technology to create trust communities? What kind of business and political decisions affected the public's access to this technology? Finally, does the trust community evolve into an institution—either joining or replacing the establishment they challenged? The chapter includes the six cases shown in table 3.1.

TABLE 3.1 Activist Cases

	New technology?	*Create a trust community?*	*Transform into an institution?*
China: Protect Guangxu (1900)	telegraph	Yes	No
Philippines: Ouster of Estrada (2001)	mobile phone	Yes	No
Taiwan: Bring DPP to power (1980s–1990s)	cable television	Yes	Yes
International Campaign to Ban Landmines (1990s)	Internet	Yes	Yes
Egypt (2011)	Facebook, YouTube, Twitter	Yes	To be determined
Tunisia (2011)	Facebook, YouTube, Twitter	Yes	To be determined

China 1900: Protecting the Emperor with Public Telegrams

In 1900, Empress Dowager Cixi placed under house arrest Emperor Guangxu, leader of the "Hundred Days Reform." Guangxu called for major changes in the Chinese government. He supported practical training for officials that went beyond the classical, literary training they had been given. He thought science should be added to the college curricula, including fields like medicine, mining, industry, and railways. He wanted local governments to support commerce, industry, agriculture, and exports. Also, he argued that the military budget should be spent on modern equipment, not on the redecoration of the empress dowager's summer palace.[1]

To create public pressure on the empress dowager not to remove Emperor Guangxu, Jing Yuanshan, chief of the Shanghai Telegraph Administration, sent public telegrams to newspapers around China and around the world, which in turn triggered more telegrams in support of the emperor from within China and abroad. These telegrams, one of which was signed by over 1,200 people, appealed to the empress dowager not to depose Emperor Guangxu. In the end, Empress Dowager Cixi, surprised by the vehemence of public opinion, was not able to replace Emperor Guangxu. However, Jing Yuanshan fled, was later arrested in Macau, and was released after a year and a half.[2]

As China's center of commerce, Shanghai was also the center of the newspapers that were critical to the success of the public telegrams. Most newspapers in Shanghai supported Guangxu; only one sided with the empress dowager. Telegrams sent to Shanghai had an easy time getting further circulation.[3] With the advent of public telegrams, instead of taking weeks to get information, newspapers were able to gather news in two or three days. Newspapers that printed public telegrams attracted more readers; those that did not, failed.[4]

The Qing court sought to limit and control the telegraph from the start. Until 1894, in principle the government controlled the telegraph to maintain control over the country. However, in practice it was difficult for the government to control traffic, especially over foreign-controlled cables that carried messages to other countries. Beginning in 1895, activists began using the telegraph for public telegrams to express political opinions. Especially after 1905, telegrams were widely used to mobilize support for political causes.[5]

The state-owned telegraph, which paralleled commercial lines, was not profitable. In contrast, Jing's network was. He lengthened service hours and dropped the price of telegrams. He built a line between Shanghai and Guangzhou, still China's main economic corridor today. From 1884 to 1898, his profits increased five-fold.

Jing leveraged his business success for social influence. He successfully raised humanitarian funds after a series of natural disasters hit China in the late 1870s. The Qing court commended him for his work.[6] However, he also failed. In an earlier public telegram campaign, Jing raised funds for a volunteer army to fight Japan and for a special codebook for correspondence. But China negotiated a peace treaty with Japan before Jing's ideas got off the ground.[7]

Things were different in 1900 when Jing sent his public telegram in support of Emperor Guangxu. This time others also joined. In Shanghai, a group of several hundred Christians and merchants sent a protest telegram to Beijing. In the southern city of Guangdong, another few hundred gentry came together to send a telegram. In the eastern city of Hangzhou, thousands signed a telegram sent to Beijing. Chinese communities in other countries also sent telegrams. Messages come from Thailand, Singapore, Malaysia, Hong Kong, the United States and Japan. Later in 1900, on the occasion of Empress Dowager Cixi's birthday in October, several groups from Southeast Asia repeated their concerns as they delivered birthday greetings.[8] In this case, the telegraph facilitated the creation of an alternate trust community with ideas that challenged the establishment.

Philippines 2001: Phones, People Power, and Ousting of the President

"As long as you[r cell phone] is not low on battery, you are in the groove, in a fighting *mood,*" quips columnist Mahou Manghas.[9] On the morning of Tuesday, January 16, 2001, the Philippine legislature voted to exclude potentially incriminating evidence from a trial in which President Joseph Estrada stood accused of taking payoffs from illegal gambling operators. People heard the news on television and radio and erupted in protest. They massed spontaneously in the streets, following notices spread by Short Messaging Service (SMS) on mobile phones, demanding the resignation of Estrada. By midnight there were thousands of people gathered along the Epifanio de los Santos Avenue, also known as the EDSA highway, the historic site of the "people power" protest that ousted President Ferdinand Marcos in 1986. SMS kept protestors coordinated, updated the news, and managed food delivery. Activists preferred SMS, which were more difficult to trace than landline phones. During the week of the uprising, around 70 million SMS were sent in a city with less than 2 million inhabitants.[10] By Saturday, January 21, Estrada was forced out, and a new president, Gloria Macapagal-Arroyo, was sworn in.[11]

From his inauguration as president in 1998, Estrada had intimidated the media. Since direct muzzling was not politically acceptable, he organized a special office to bribe journalists for favorable coverage. Estrada's operation was on a greater scale than was typical for the Philippines. Reportedly, his monthly budget was US$40,000, drawn from payoffs from illegal gambling operators. In addition, Estrada threatened newspapers that criticized him with tax audits. The *Manila Times* shut down in 1999 after such threats. Against another critical paper, the *Philippine Daily Inquirer,* he persuaded movie producers to not place ads.[12]

To get real news about Estrada, people turned to networks run by civil society organizations, including the Roman Catholic Church. They used email, Internet, and SMS to spread information and jokes lampooning the Estrada administration. Traditional media picked up on the growing rebelliousness in the informal media. In 1999, when the *Pinoy Times* started covering Estrada's personal life, his mistresses, and his growing personal wealth, circulation rocketed from a few thousand to several hundred thousand. By 2000, when news surfaced about Estrada's intrigue with illegal gambling operations, people were primed.[13] Ironically, Estrada's heavy-handed controls fostered alternate news networks that were so trusted in a crisis that they could trigger his overthrow.

The Estrada episode occurred at a special moment in the Philippines' mobile phone development. Figure 3.1 shows the growth of fixed and mobile phone subscribers and the growth of Internet subscribers and Internet users. In the late 1990s, mobile phones began to grow spectacularly. In 1999, two years before Estrada's ouster, mobile phones exceeded landline phones on a per capita basis for the first time. By 2000, there were more than twice as many mobile phones per person than fixed phones.

Mobile phones grew in part because of key political and economic policies. Since the 1920s, PLDT, the Philippines Long Distance Telephone, had been the monopoly phone company. In 1992, the World Bank identified this monopoly and its poor telecom service as a major bottleneck to the country's economic development. Singapore

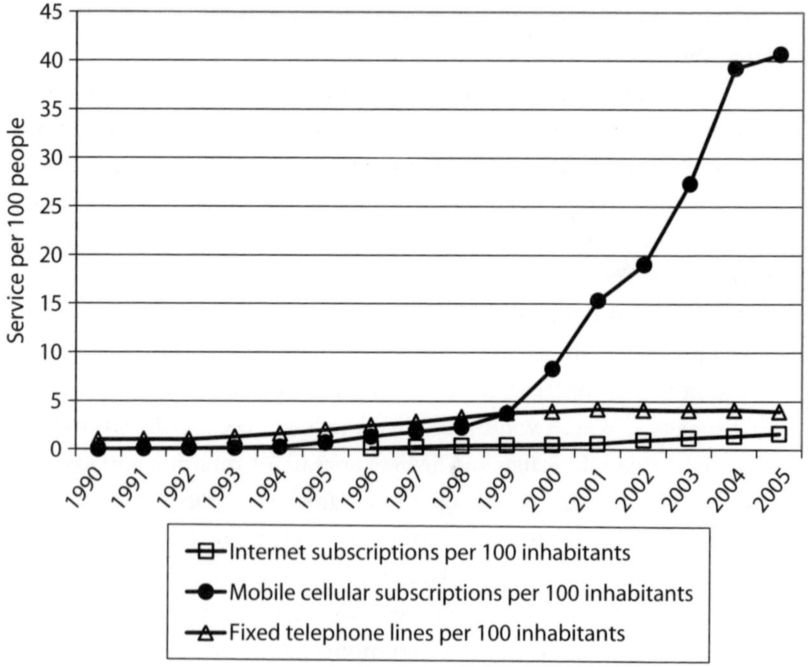

FIGURE 3.1. Philippines Communications Services, 1990-2009.
Source: International Telecommunications Union.

Prime Minister Lee Kwan Yew pointed out, "The joke in the Philippines is that 98 percent of the population are waiting for a telephone, and the other 2 percent for a dial tone."[14]

In 1993, Philippine President Fidel Ramos issued Executive Order 109, which broke PLDT's monopoly and allowed any firm to provide international or mobile phone service.[15] Telecom was the lead sector in Ramos's economy-wide anti-monopoly campaign. While the build out of landline service was mixed, mobile phone service leaped ahead. In 1992, mobile covered only 20 percent of the country; by 1998, 37 percent of the country was covered, with more than a dozen operators competing to offer service.[16] Filipinos loved texting on mobile phones. The subscribership of Globe Telecom, the first company to offer SMS, surged from 420,000 in mid-1999 to 725,000 by year-end. Its most expensive SMS package was an affordable US$4 (150 pesos) per month.[17]

After SMS helped install her as president after Estrada, Gloria Macapagal-Arroyo (GMA) established TXTGMA, a service to allow Filipinos to text their concerns to her. She received messages that ranged from reports of illegal drugs in a small town, the condition of Filipino workers abroad, and bad traffic in Manila, to complaints about government officials. Arroyo also used the messages to gauge public opinion on issues such as the suspension of the death penalty.[18] However, people used SMS against her as well. In 2005, when Arroyo was accused of manipulating an election, SMS political jokes circulated quickly. "GMA wants her SONA [State of the Nation Address] but the public just wants her GONA [Gone *Na* or Gone Already]." By 2005 the political use of cell phones advanced from texting to include ringtones. The sound file of the allegedly incriminating conversation between Arroyo and election official Virgilio Garcilliano was turned into a ringtone called "Hello, Garci." Within a day of its creation, the ringtone site reported 48,000 hits before it crashed. Within three weeks of the release of the audio conversation, there were more than twenty versions of the "Hello, Garci" ringtone.[19]

Taiwan 1970s: Cable Television and the Rise of Democracy

Prior to the 1990s, the signature policy of Taiwan's authoritarian government was that in the future it would return to govern mainland China, power it had abandoned when the Nationalist Party lost to the Community Party after a civil war. In contrast, others thought Taiwan should become an independent nation. The first priority of opposition politics in Taiwan, however, was transforming the authoritarian government into a democratic one.

In Taiwan, the government controlled television from the start in the 1960s.[20] In the mid-1970s, cable operators illegally installed videocassette recorders, coaxial cable, and transmission equipment.[21] Since there were three main government-controlled stations, cable television became collectively known as the "fourth channel." Cable boomed because people wanted more news, information, and entertainment.[22] Given its history of providing an alternative to government broadcasting, illegal cable television in Taiwan was closely identified with opposition politics.

In parallel with the fourth channel, the opposition Democratic Progressive Party (DPP)'s "Green Team" created reports on topics like the condition of workers and fishing communities that provided an alternative to official news broadcasts. These illegal reports were distributed by videocassette and played on cable television.

In 1989 the DPP was first allowed legally to run candidates, but its application for a television station license was rejected. Instead, they established a guerilla "Green Television Station." Using smuggled equipment, they broadcast a two-hour program to introduce its candidates to the electorate two days before the election.[23] The government ordered a full-scale but ultimately ineffective crackdown. Censors would cut television cables in the morning; cable operators would reattach them in the evening.[24]

For many years, the illegal cable television system provided Hong Kong soap operas, Japanese variety shows, pirated movies, stock market information, travel, religious, and other programming unavailable on terrestrial television. Households paid US$11–22 a month for a package of thirty to forty channels with around-the-clock programming.[25] Between 1990 and 1995, cable television grew from about 11 percent to 54 percent of all households. This growth took place while the government lifted martial law in 1987 and sedition law in 1992. All media, including cable television, expanded in numbers, diversity, and willingness to challenge government news.[26]

In 1990 the DPP founded the Taiwan Democratic Cable Television Association, a group of about fifty cable systems that broadcast news and information from the opposition perspective. They jointly produced political programs. The DPP protected members from government crackdowns and sought legalization. The association also initiated a Congressional Channel that aired the proceedings of the Legislative Yuan and National Assembly, unedited and without commentary.[27] In the meantime, a second opposition party, which had splintered from the Nationalists, entered the market and also sought to legalize cable television. Another factor was the international movie industry, which pressed the government to end the illegal pirating of their content by local cable operators.[28]

In 1993 the ruling Nationalist party, faced with pressure from two opposition parties and foreign investors, passed the Cable Law, which in effect legalized the expression of opposition views on television.[29] After 1993, several all-news channels focusing only on Taiwan politics were created. Political call-in shows were one of the most popular program formats, especially during election campaigns.[30] In 1996, in this transformed media environment, Taiwan held its first presidential election and became a full democratic government.

The Green Team videos and the Green Television Station created an atmosphere in which people became more comfortable openly articulating an opposition point of view. The cable television associations were institutions that destabilized the authoritarian state. Making cable legal opened political debate on Taiwan's television and contributed to the evolution of its democracy.

Global 1990s: Banning Landmines with an Information Landslide

In just five short years, a few nongovernmental organizations (NGOs) persuaded nations to sign an international treaty to ban landmines. In the early 1990s, the 100 million plus landmines in more than sixty countries injured or killed more than 2,000 civilians per month. In the previous fifty years, landmines killed more than nuclear and chemical weapons combined.[31] In the early 1990s, more than one hundred companies and government organizations manufactured over 340 types of antipersonnel landmines. While every year 80,000 were cleared, 2.5 million more were planted.[32] The International Campaign to Ban Landmines (ICBL) was launched in 1992, bringing together over 1,400 NGOs in more than ninety countries. The campaign called for an immediate, permanent ban on antipersonnel landmines and commitments to remove existing mines and assist the victims of mines. In 1997, a treaty was finalized. Over 120 countries signed it within fourteen months.[33]

The ICBL began with a fax[34] from Bobby Muller from the Vietnam Veterans of America Foundation (VVAF) in the United States to Thomas Gebauer of medico international in Germany. In 1991, Muller asked Gebauer for help with a victim assistance project; soon after, they called for an international campaign. In 1992, six major NGOs came together to establish key goals.[35] In 1993, the ICBL held first NGO conference. Jody Williams of the VVAF was elected coordinator. Williams and her colleagues were later awarded the Nobel Peace Prize for their accomplishments.[36]

Williams emphasizes that clear and consistent communications were critical to the success of the campaign. Phone, fax, and email were essential to internal coalition communication; face-to-face communication was essential in communicating with governments and militaries.[37] The coalition's view is that the campaign drew momentum from the network of people, not the network of technology. As figure 3.2 demonstrates, between the launching of the coalition in 1992 and the signing of the treaty in 1997, Internet use and cell phone use in high-income countries grew dramatically. The early leaders of the campaign hailed from these regions. World averages rose later and more slowly. To reach partners in the field, the campaign relied not just on email but on a variety of communication methods. Like the story of Jing Yuanshan and the public telegrams in nineteenth-century China, the latest in technology carried information quickly from center to center, and older communications methods—meetings, newspapers in the case of Jing, meetings, phones, and fax machines in the case of ICBL—maximized the distribution.

The ICBL coalition grasped that technology gave it an advantage even over nation states. Indeed, Williams recollects, "The ICBL has often learned of developments relating to the ban movement before governments became aware of them. This has made the ICBL a focal point of information for governments and non-government organizations alike. Its role as an information center also helped build

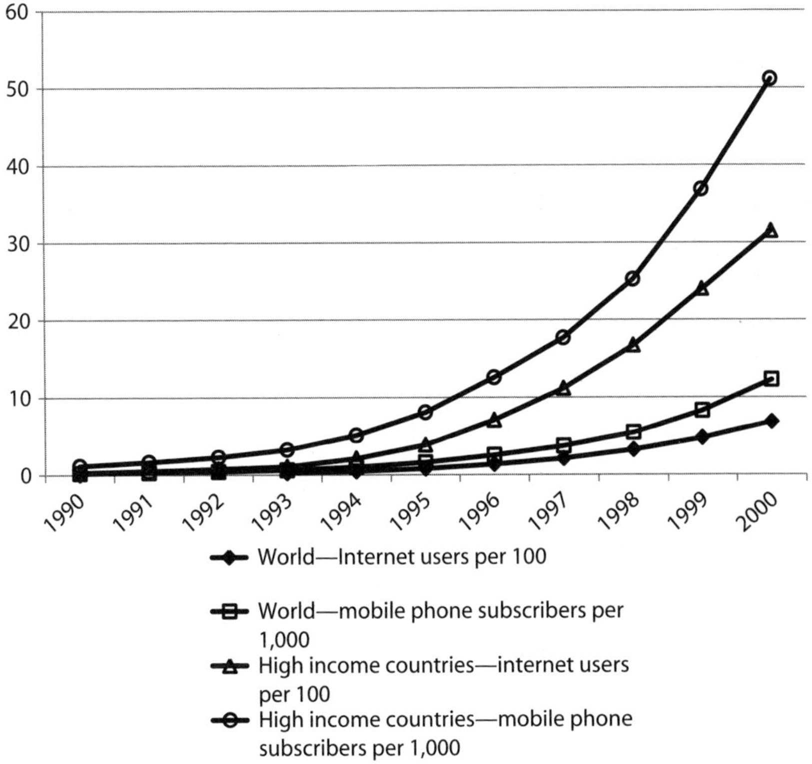

FIGURE 3.2. Availability of Internet and mobile phones, 1990–2000.
Source: World Bank Development Indicators.

confidence between governments and the ICBL."[38] In other words, the ICBL became the center of this trust community. Its network was so successful in bringing together information that it became an institution as significant as national governments. ICBL's use of a range of communications technologies set it apart from other NGOs. While some think of the ICBL as an example of the new global civil society, a transnational organization able to challenge powerful nation states, in fact the main leaders were from the wealthy world with technology access.[39] ICBL was effective, however, because they had direct connections to economic developing regions plagued with landmines. Without this connection, the campaign would have failed.[40]

After the treaty was signed, ICBL's power base was its network of people who could quickly gather information on landmine activity around the world. When the Mine Ban Treaty was signed in 1997, the ICBL achieved its purpose and its future was unclear. The treaty did not include any provision for tracking compliance with its terms; some states believed further institutionalization was unnecessary. Concerned that the sense of crisis would deflate, in February 1998 the ICBL refocused its efforts to implementing it by monitoring members' progress. In June 1998, they created the Landmine Monitor Group that consisted of five NGOs led by Human Rights Watch.[41]

Mary Wareham of Human Rights Watch recounts that in the subsequent months the Land Monitor Group produced a report on every country in the treaty, including current stockpiles, humanitarian action, and assistance for survivors and funding for programs. By May 1999, less than a year after the Landmine Monitor Group was created, they distributed a report of 1,000-plus pages, including thematic assessments across countries, to ICBL ambassadors at the first meetings of states, held in Mozambique.[42] Wareham describes the scene:

> The room fell silent as diplomats flipped through the pages to their country entry—each one curious to see how their country was portrayed on the landmine issue. The cover featured a youth running swiftly on his prosthetic leg, conveying a positive message that worked well with the report's subtitle, "Toward a Mine-Free World." Many delegates were surprised to see the report. . . . According to [Steve] Goose [ICBL campaigner from Human Rights Watch], "It stunned people. I don't think they had any notion that we were going to be able to pull together something that had so much information. The length itself shocked everybody, and there was no 'filler' in it; it was very dense and filled with facts. People began to realize right away too that it wasn't a 'polemic.' It was in fact a very factually based approach to gathering information, a baseline for information from which to gauge progress."[43]

Once again, the ICBL capitalized on its ability to collect credible information at a speed and quality greater than nation states. The reports are issued annually. The Landmine Monitor tracks members' reports on the actions they take to adhere to the treaty. The pressure of the publicity from the Landmine Monitor has contributed to greater compliance with the treaty's requirement that states file reports. In 2001, only 63 percent of states filed; in 2006, 96 percent of states filed. Often the state's reports to the treaty regime include material from the Landmine Monitor.[44] More than ten years later, the 1999 Landmine Monitor report and every subsequent annual report is available for free on the Internet.[45]

Egypt and Tunisia 2011: History, Social Media, and Revolution

The Egyptian and Tunisian governments toppled in 2011 because longstanding frustrations were set alight by catalytic events. In Egypt, modern youth political involvement dates back to the 1970s when such activism was banned in the schools. There had been groups struggling against Mubarak for years.[46] In Tunisia, economic growth and development suffered for years under the Ben Ali regime.[47]

In Egypt, one catalytic event was the beating death of blogger Khaled Said by police in June 2010 for exposing their corruption on his blog. Wael Ghonim, a Google worker, started a Facebook group to commemorate him. A Facebook group allows people to share messages, pictures, and videos online with all registered users of Facebook, with special notice to those who sign up for the group. This Facebook page became a logistical tool for the protest community. Ghonim also used his Twitter account to link together a large network of people both inside and outside of Egypt.[48]

In parallel with events in Egypt, on December 17, 2010, in Tunisia, a street vendor named Mohamed Bouazizi set himself on fire after his failed appeals to the police, town government, and regional governors to fight an inspector's fine. News of his case spread through blogs and text messages on cell phones even though the state-owned media did not cover the incident. On January 4, 2011, Bouazizi died. Bouazizi's plight ignited frustrations building among Tunisians who were angry at the deepening corruption of Tunisia's leader, Ben Ali, and his family. The people protested, and by January 14, 2011, Ben Ali fled Tunisia.[49] On January 25, 2011, in Egypt protestors began appearing in Tahrir Square in Cairo, the central dramatic location of the demonstrations that toppled the Mubarak regime on February 11, 2011.

The social media involved in these events are new: blogs have been online only since 1997; Facebook, since 2004; YouTube, since 2005; and Twitter, since 2006. With only twenty years' history, their impact on creating trust communities is just beginning. A blog is a web log, a diary of events chronicling the author's views, self-published online. Blogging platforms in this period are often free to bloggers; the commercial platforms that support them generate revenue from advertising. Internet search engines are online tools that allow users to search for content on the Internet. These are also free to users since search engine companies generate revenue from advertising. In 2011, Google was the most prominent Internet search engine worldwide, although in specific country markets there may be other engines that are more popular. Google also owns Blogger, one of the largest blogging platforms, and the largest video platform, YouTube.

Facebook is a social media application. Users can create their own "page" in this online "book" on which they can post pictures, messages, videos, and links to other sites. For users who are Facebook "friends," their Facebook pages are linked, and postings on one friend's page will appear on all his or her friend's pages as well—a kind of personal broadcasting station. Facebook is also a commercial platform, with most of its revenue generated from advertising. Facebook users receive their pages for free.

Twitter is a social media application known as a microblog. Like blogging, each user has a page to chronicle happenings and messages. In contrast with blogging, the messages on microblogs are often very short—in the case of Twitter, less than 140 characters per message. "Followers" of a particular user's Twitter account automatically receive all messages posted by that user. Twitter is also a commercial platform with revenue generated from advertising; users pay nothing to open a Twitter account. YouTube is a social media application that allows users to create their own media "channel," like a television channel, and post videos that are visible to the public or to selected friends. Users pay no fees to open a YouTube account; YouTube generates revenue from advertising. Google, Facebook, Twitter, and YouTube are commercial, not state-owned companies, and all are headquartered in the United States.

All these applications require users to have access to the Internet. In both Egypt and Tunisia, the telecommunications network that gives people access to the Internet has a history of state control. As of December 2010, state-owned Tunicell had just over 45 percent of the cell phone subscribers in the country, and state-owned Tunisie Telecom had 70 percent of all broadband subscribers.[50] Nevertheless, activists could access communications networks that governments could not easily suppress. Cell phone, texting, and Internet service continued. When protestors used Facebook, Twitter, and YouTube to share news and messages, governments could not completely block it. Activists were able to circumvent government blocks, websites resisted government hacks, and hackers attacked the government sites.[51] In 2011, one-third of Tunisians had Internet access; among university students, most visited Facebook at least once a day, on average spending nearly two hours on the site; and two-thirds relied on it for information about the demonstrations.[52]

In Egypt, Cairo is a major regional telecommunications and media hub in the Middle East region. While all the major cellular phone providers are privately owned, state-owned operator Telecom Egypt had 60 percent of all broadband subscribers as of December 2010. As protests grew in January 2011, Mubarak tried to suppress them by shutting down part of Egypt's Internet to cut off the flow of information. The first shutdown order was issued on January 28, 2011, but compliance was uneven. Service was interrupted for several days. The interruption did not discriminate—government offices as well as activist communications were affected.[53] Figure 3.3 shows that these

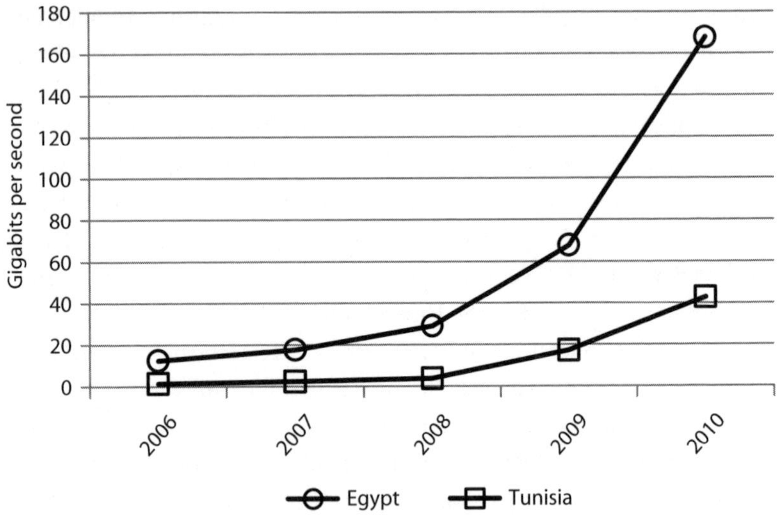

FIGURE 3.3. International bandwidth usage for Middle Eastern and North African countries, 2006–2010 (gbps).
Source: Telegeography, "Global Bandwidth Research Service: Middle East and Africa."

events took place at a moment of sharp growth in Internet capacity. From 2008 to 2010, the physical capacity of the infrastructure underlying the Internet grew at least 500 percent for both Egypt and Tunisia. As the unrest unfolded, those who had access to the Internet had much better access than before.

While social media have been the focus of most of the Arab Spring discussion, blogging—an older Internet form—was still important. It was the death of blogger Khaled Said that set off protests in Egypt. In general, blogging is well-established in the Middle East as a vehicle for personal and political expression.[54] In earlier protests such as the 2004 activity protesting the war in Iraq, blogs were used to build networks of activists. Bloggers often include in their postings links to television news. Etling's 2009 study of the Arab blogosphere records that Al-Jazeera and British Broadcasting Corporation (BBC) were the third and fourth most-linked to sites on Arab blogs. The first and second most-linked sites were YouTube and Wikipedia, which both have content generated by members. Al-Jazeera's television reporters and talk show hosts use Twitter and other social media to augment their audiences. Around this time Wikileaks' release of reporting cables by the American ambassador confirmed Tunisians' own suspicions of the venality of the Ben Ali's corruption.[55]

In Egypt and Tunisia, the 2010–2011 uprising demonstrated the skills activists had acquired from past organizing. One Egyptian political activist recounts the role of the Internet in the Youth Movement in 2004–2006, when blogs were used to spread ideas. In 2008 the Youth for Change movement used Facebook to organize a strike in which 50 percent of Egyptian workers participated.[56] In Tunisia, the number of civil society organizations grew from around 2,000 to 10,000 between 1988 and 2011. Organizations like the women's associations that developed during this period were well represented in the 2010–2011 uprising. In both Egypt and Tunisia, it remains to be seen how the new skills acquired in the most recent collective actions are retained and re-applied in new challenges and whether the information centers that held so much social capital will grow into institutions with regularized routines of work and communication.

Conclusion

Throughout history, activists have often taken advantage of the latest communications technologies to share ideas and information. They express a different vision of how the world could be. In the process, they create networks of people that coalesce into trust communities, some transient, others permanent.

Catalyst. In all the examples, an idea pre-existed the movement, but a catalytic event created momentum for change. Once an opinion was exposed, there was a groundswell of support. Technology does not cause this to occur; it enables news of it to travel faster. People withhold expressing ideas if they think they are alone. Once they realize they are not, they feel free to reveal these ideas; information cascades as more join in.[57] In nineteenth-century China, Jing Yuanshan supported a number of causes, some of them with momentum and others without. In the Philippines, there were a range of problems with Estrada's administration, but news of the kickbacks from illegal gambling operators triggered his downfall. Once the idea breaks through, there is an opportunity for people to newly identify themselves with it.

Building trust through work. In all cases, a new set of work skills—a repertoire to organize and express views—developed alongside the new technology. The collection and distribution of information in these cases, like the *TsunamiHelp* volunteers in chapter 2, was the core work of the movement. In the case of ICBL, the core work of collecting and sharing information on mine activity was the foundation for the monitoring group that continues today. Relationships and trust built up around the ritual of collecting and distributing information hold communities together. The work of communicating creates opportunities for reciprocity—whether sharing personal stories, expressing political opinions, or receiving tactical instructions.

Technology change. New technology expanded the resources available to activists. This was true for the telegraph, the cell phone, cable television, and the Internet. A group that once took weeks to mobilize by word of mouth can now be mobilized in days by cell phone. Information that once could be delivered only by airmail, like the *Landmine Monitor,* can now be distributed for free, instantly, online.

Business. In all these cases, the key technologies were commercial, built not by state networks but by private business. Jing's telegraph in China was a business success. In the Philippines, short messaging service was a killer app for the mobile phone operators. Cable television in Taiwan spread because people wanted entertainment, not just news. The social media tools used in Egypt and Tunisia in 2010–2011 were products of international companies inclined to resist efforts by national governments to interfere in their business activities. This underscores the close connection between economy and politics.

Old and new media forms. The new technology was not used in isolation, but in conjunction with older media. In China, the telegrams were reprinted in newspapers. In the Philippines, in addition to SMS on mobile phones, newspapers, television and radio also kept people informed. In Taiwan, the Green Team circulated videotapes before they used cable television; moreover, media liberalization included not only the development of cable but also the rise of magazines and newspapers. For the ICBL, the organizers used not only email but also faxes, phones, and face-to-face communications to collect and distribute information. Rarely does a new technology cause an earlier one to disappear; instead, the old technology adapts to specialized uses.[58] While people fly planes, ride trains, and drive cars, there is still a place for cycling and for walking. The same is true for communications technologies.

Information as power. These cases illustrate that in politics, information is a resource, a type of currency, a class of ammunition. Especially for activists, their communities are built around the cache of ideas and information that informs their common worldview. The ICBL used the information they collected on landmine developments to hold government accountable. Information cascades broke spirals of silence. In Taiwan, the cable television networks provided an alternative view of politics to that based on the information on the state-run news programs. In the case of Taiwan and ICBL, these repertoires were accompanied by the rise of institutions with specific goals and ability to negotiate with other political institutions. In Taiwan, these were the cable television associations. In ICBL it was the formation of a new nongovernmental organization which has expanded its remit to include other issues. In all these cases, activist networks provided ideas and information that challenged the government view. Of course, governments have their tools to express their vision and influence infrastructure, information, and ideas. The next chapter discusses them.

CHAPTER 4

Governments Shape Nations with Communications Technology

"CANADIANS HAVE NEVER been united as they are on July 1, 1927," opines Canadian Broadcast Corporation in its online archive, where it is still possible to hear Canada's first nationwide broadcast on its Diamond Jubilee, celebrating sixty years as a nation. "Never before has there been such an attempt at globe-circling broadcasting as that which is being participated in today and tonight," we can hear Senator George P. Graham say. Along with the speeches, there is the national anthem, "O Canada," travelling to us in the twenty-first century from 1927, crackling with the imperfections of radio broadcasting and sound recordings.[1]

To exercise power, governments must communicate. In fact, without the ability to communicate, governments cannot exist. National governments are acutely aware of the intimate connection between their ability to govern, the quality of the communications network, and the ideas that flow over it. In most countries, governments built the first telephone and telegraph networks. Often they ran the first radio and television networks. In many countries, the biggest Internet service provider is a unit of the largest telephone network, often state-owned. In the previous chapter, I described how activists used information as their source of power to challenge the state's worldview. This chapter focuses on that worldview and how governments work to maintain it. It also shows that governments often fail to persuade people to accept their point of view. The interactive nature of communications means that the message sent might not be the message received; instead it may be a message reinterpreted.

Governments have several goals in setting policy for communications networks and services: protecting national security, fostering a national identity, encouraging economic development, and improving delivery of public services. National security encompasses defending the nation against foreign attack, maintaining domestic order, and promoting domestic safety. National identity includes creating, maintaining, and defending a national culture and projecting it to the world. It involves including or excluding people as part of the nation. Economic development includes promoting infrastructure and services as a strategy for better growth and productivity.

Two aspects of communications networks are the infrastructure and the information that rides on top of the infrastructure. The next sections explore the relevance of these priorities, first for infrastructure policy and then for information and media policy. Government can either extend or constrain the growth of infrastructure and the spread of information. Sometimes they succeed, sometimes not.

Many theorists have worked on the conceptual relationship between communications networks and the idea of a nation. Benedict Anderson (1991) called nations "imagined communities." People of the same nationality feel a kinship with each other, a kinship that may reference a geographic place but is not bound by it. Fellow citizens abroad are still citizens. Nations are imagined because any individual in a nation is unlikely to have the opportunity to meet every other individual: our connection is not face-to-face but mediated in some way—by newspapers, by television, or some other means. Before Anderson, Karl Deutsch (1966) focused his attention on communication's mediating role and its connection to nationhood. Everything that holds a nation together—a common culture, the ability to act together, the exchange of ideas and information, and economic cohesion—requires the infrastructure and facilities to communicate. He underscores not only the idea of nationhood, but also the communications among people that are required to circulate ideas that foster this common consciousness.

After Anderson, Manfred Steger (2008) takes the idea of an imagined community one step further. He reflects on the development of "global imaginaries." Like Anderson and Deutsch, Steger's imaginaries are also communities—communities held together by ideas and the means to communicate them. In a time when the technology allows richer communication across national borders, he proposes the idea that communities larger than the nation are now possible and emerging. The globalization of life is the basis of Steger's thesis that the global imaginary may supplant the national one. This possibility is the basis for a range of government policies and institutions that have developed over a century and a half to defend the idea of a nation. How easily nationhood can be challenged remains to be seen.

This chapter includes several cases from book publishing in Russia in the seventeenth century to satellite television in the Middle East in the twenty-first century. Table 4.1 provides a quick summary of the cases and the main government policy objectives they illustrate. The chapter first discusses government approaches to infrastructure, then approaches to information and ideas. Within each

TABLE 4.1 Government Case Studies

Country	*Dates*	*Case*
Canada	1927	Telephone network ties the nation together
Brazil	1900s	Telegraph network reaches into the Amazon
China	1980s	Telegram growth and tensions with national security
US	1860s	Telegraph helps Union win the Civil War
US	1980s	Internet commercialization prioritized
USSR	1960s	Information control prioritized
Russia	1880s–1917	Print publishing censored by the Tsarist regimes
WHO	1990s–2010s	World Health Organization improves data collection on disease
US	1960s–2010s	Internal Revenue Service improves electronic filing of returns
India	1987	TV *Ramayana* influences political vocabulary and iconography
UK	1938–2004	BBC becomes empire's voice to the world
Qatar	1996	Al-Jazeera raises Qatar's diplomatic profile

section, I discuss national policies aimed at promoting national identity, preserving national security, and enabling economic growth.

Infrastructure and National Identity

CANADA 1927: TELEPHONE TIES THE NATION TOGETHER

To celebrate its 1927 Diamond Jubilee, the sixtieth anniversary of its founding, the Canadian nation asked an ordinary favor from its communications industry—to deliver a message from the capital to all of its people. To mark the national celebration, Prime Minister William Lyon Mackenzie asked radio stations, telegraph, and telephone companies' cooperation to make the first nationwide broadcast. To execute this request, radio stations across the country were linked by wires, both telecom and telegraph. Since telecom networks had spotty coverage compared to telegraph, companies were unsure of its commercial viability and were reluctant to invest. However, joined in their rivalry against the telegraph, the telephone companies collaborated on the Trans-Canada Telephone System. Four years later, in August 1931, it was inaugurated in a symbolic ceremony led by the governor general, just weeks after Canada won its first international legal status as a country with the Statute of Westminster. From then on, the telephone was tied to the strength to the vibrancy of the Canadian nation.[2] Canada's east-west infrastructures tie its provinces together in counterbalance to the north-south systems binding the United States and Canada.[3]

BRAZIL 1900: TELEGRAPH REACHES THE AMAZON

In 1938 when French anthropologist Claude Lévi-Strauss visited Brazil, he said the telegraph network, broken and abandoned, "heightens the loneliness of the region, rather than reducing it." In Brazil, a military-led expedition to unite the country by telegraph foundered without support from commercial demand for the network.[4] How did this happen?

In his research, Todd Diacon discusses how Brazil in the nineteenth century was still a new country, and it wanted to be more united. Like Canada, Brazil has scale; it is the world's fifth largest country by geographic expanse, 2,500 miles from north to south, 2,700 miles at the widest point between east and west. A challenge in the nineteenth century was whether Brazil really was a single nation with common beliefs and a shared vision for the future. Many government initiatives sought to unify the nation. There were public health campaigns not only to improve the people's quality of life but also to prove that ignorance, not racial inferiority, was the cause of illness and disease. A mandatory military service not only staffed the armed forces but also was a tool of national integration, incorporating people from different groups and improving the health of the poor. In this context, the government launched its campaign to build a national telegraph network.[5]

The man commissioned to build the network, Candido Mariano da Silva Rondon, is best known in Brazil for his leadership of the Brazilian Indian Protection Service, through which he preached that Indians should be treated with dignity as Brazilians, but by relinquishing their own language, culture, and community. Between 1900 and 1906, Rondon's commission built nearly 1,100 miles of telegraph lines (220 across the swamps of the Pantanal, 150 through forest), 16 telegraph stations, and 32 bridges. Rondon also mapped areas for the first time, another exercise in establishing national identity. After 1906, for the next eight years, Rondon's commission built another 1,100 miles of telegraph line and constructed around twenty telegraph stations and other necessary infrastructure—bridges, corrals, and ferries. The entire network was inaugurated in January 1915.[6]

The military staffed the telegraph service with poor Brazilians unable to avoid the required military service. Often they were illiterate, in poor health, and with criminal records. The military often sent the most rebellious and troublesome of soldiers to the telegraph commission. The commission soldiers along the telegraph line suffered from poor living conditions and illness, especially malaria, the greatest health threat to living in the Amazon. Attacks from Brazil's indigenous people also endangered them. Furthermore, Rondon was a fierce disciplinarian. In 1912, halfway through the commission, he estimated that 10 percent of his soldiers deserted.[7]

Rondon maintained that the benefit of the telegraph was not only military but also in building a nation and developing its economy. However, his contemporary critics pointed out that a radiotelegraph that did not require construction of lines might be a cheaper and more appropriate technology to use in the Amazon. In terms of commercial development, very few commercial messages traversed the network. Most telegrams were the internal communications of the commission itself. Six years into operation, one-fifth of the messages were private; the rest of the telegrams were official. The people living around the telegraph stations were not using them, and Diacon's research shows that in the first decade of operation, revenues along some of the telegraph lines declined. In early republican Brazil, government fiat alone was unable to sustain the network marvel of its day. Gradually, stations were abandoned and lines closed down.[8]

DISCUSSION: CONTRASTING BRAZIL AND CANADA

Brazil's difficulties in maintaining a telegraphic connection from its metropolitan centers to its interior were not unique. Toward the end of the nineteenth century, there was a boom in building telegraph cable networks between countries. Britain had connected to Europe and turned to connecting to Australia, Canada, New Zealand, and South Africa. The networks followed trade and finance. Extensive commerce between Britain and South America, including Brazil, meant the international telegraph systems connected those along the coast to the rest of the world.[9] The Brazilian effort to extend that telegraph inland was a national response to a global trend. However, the international telegraph system largely followed commercial demand. When telegraph systems were expanded to less-commercial areas, governments had to subsidize their development. The same turned out to be true for Brazil's telegraph to the interior. There was no broad community of people to imbue the telegraph with meaning, despite the military and government's effort to engineer a national trust community.

In Canada, however, the telephone companies were commercial enterprises from the beginning. While some were later bought by local governments, at the start they were financially independent. They physically tied their growth to the railways across the country. In the early twentieth century the telephone was primarily a business tool used to coordinate the distribution of goods through trains and other transport.[10]

The cases of Canada and Brazil highlight the concern of governments in using communications networks to tie together regions of their nations. Governments frequently fund universal service programs that subsidize the construction of infrastructure or the price of service for people who would otherwise not be connected—rural areas, low-income groups, schools, and libraries. The objective is partly economic—to contribute to overall growth and improve opportunities for work and income—but just as important is the goal of including people in the mainstream of society however that might be defined. The efforts of Brazil to expand its telegraph network and of Canada to establish its telephone network are the antecedents to today's programs to bring Internet service to community centers or build fiber optic infrastructure to remote geographic regions.

Infrastructure, Economic Development, and National Security

"To stimulate sustainable economic growth, increase productivity, improve public services, promote transparency and good governance, enhance social inclusion, and ultimately reduce poverty" are the World Bank's announced objectives for information and communications technology projects around the world. These goals are typical of many countries and development organizations. The World Bank seeks to accelerate participation of developing countries in the global economy, increase competition and private investment in this sector, and foster economic and social development that is sustainable.[11] While in its mission there are oblique references to the usefulness of good information for states to govern well and to tie the nation together through social inclusion, the practical goals are entirely focused on market and economy. Several cases follow that illustrate this approach by nations, with varying levels of success.

CHINA 1979: TELECOM GROWTH VERSUS STATE SELF-PRESERVATION

China has made economic development the primary goal of infrastructure and information development, not national identity or national security. In fact, within China, the increased flow of information resulting from improved communications infrastructure is in some respects considered a threat to national identity and security, making the measures taken to promote communications all the more striking.

In the late 1970s, the Chinese government made economic reforms a priority, the economy grew, and demand for telecommunications services outstripped supply. Every government department was responsible for organizing its own telecom service. In China, in 1979, the telecommunications ministry centralized its authority over these departmental networks. In the 1980s, as reforms deepened throughout the economy, telecommunications development became a more important priority for the government, which gave the telecommunications industry preferential tax rates, privileged rights to retain foreign exchange, and easier terms on repayment of state loans. The 1980s were also an era in which four special economic zones and fourteen coastal cities were permitted to experiment with market economy measures. By 1991 these areas accounted for nearly a quarter of the telecommunications network in China.[12] The 1990s were a period of rapid telecommunications growth in China. Notably, cellular phones using wireless telephony technology emerged during this decade as a major commercial service. In areas dense with customers, wireless networks were often cheaper and quicker to install than wireline networks. By 2001 nearly half of all telephone service subscriptions were for wireless service.

As infrastructure grew and services became popular, the Chinese government also expanded its controls over the communications network. From the beginning, the government filtered Internet content. Its system for influencing domestic Internet content through surveillance, actively engaging people online, and police action is the most sophisticated in the world. The government's primary policy goal is maintaining national security. It monitors public discussions online that criticize the government. Sometimes government operators will participate, and sometimes the space will be closed. Therefore, while the increase in communications services and information flow is positive for economic growth, the government also believes there are dangerous elements. As the communications networks and services have grown in keeping with the state's economic goals, the surveillance and control over the content has also had to develop rapidly to mitigate criticism of the government.[13]

UNITED STATES 1864: TELEGRAPH AS THE UNION'S SECRET WEAPON

"The value of the magnetic telegraph in war cannot be exaggerated, as was illustrated by the perfect concert of action between the armies in Virginia and Georgia during 1864. Hardly a day intervened when General Grant did not know the exact state of facts with me, more than fifteen hundred miles away as the wires ran," said Robert Sherman, one of President Abraham Lincoln's leading generals in the US Civil War. Lincoln, Grant, and Sherman succeeded in maintaining the United States as a single country against a technologically less sophisticated Confederate military.[14] Other than the White House, the place where President Lincoln spent most of his time was the telegraph office just next door in the War Department's library. Lincoln regularly reached into the telegraph operator's drawer and read all the dispatches that had come in since his last visit—not only the messages to him but also all the messages sent between his generals.[15] Lincoln was the first American president to deploy telegraph technology in the service of war. It was this coordinated military action between Grant and Sherman that brought the war to a close soon thereafter.[16]

Prior to the Civil War, manufacturing industries and the railroads spurred telegraph development. The telegraph enabled more centralized management of these enterprises. The telegraph similarly centralized Lincoln's management over the military during the war.[17] By contrast, in the American South, the Confederacy rebelling against the Union that Lincoln represented made scant use of the telegraph, resisting the technology by establishing laws that prevented the lines from crossing state borders in the South.[18] The commercial companies that built the telegraph to meet business demands in the American North built a network that later gave the Union trust community—the government and military—an advantage over the South. The history of communications technology is filled with stories of militaries seeking an edge against their enemies. Whether communications networks built for national security objectives succeed depends on whether they are also commercially sustainable.

UNITED STATES 1968: INTERNET PUTS COMMERCE FIRST

One hundred years later, the U.S. government's Advanced Research Projects Agency (ARPA) funded the invention of the Internet, a decentralized communications network that could withstand a military attack. From the beginning, the uses of the Internet were never exclusively to the military. Scientists shared computer capacity across research centers. Email was the killer application that drew people to the network. In the 1980s, the military transferred the ARPANET network to civilian government oversight. In the 1990s, the government officially allowed commercial use of the Internet. Once the network went global, it became the dominant communications network of our time.[19]

In the United States, the military was important to the development of the telegraph and the Internet, but the expansion of the infrastructure to reach the ordinary citizen was done commercially. While military use was the catalyst for the Internet's invention, military objectives alone did not sustain the network. The users of the Internet worldwide are the trust community that imbue with meaning this infrastructure originally built for the American scientific community.

Information, Ideas, and National Security

USSR 1960: "INTERNYET" PUTS INFORMATION CONTROL FIRST

The Soviet Union confronted similar technical challenges in the 1960s, the same period in which the U.S. military developed ARPANET, but had a sharply different response. Gerovitch (2008) describes the Soviet response to computer networking as "InterNyet." While the Soviet Union was a technologically advanced country, the state pursued a communications technology policy that restricted the distribution of equipment and controlled the development of networks in order to better separate units than connect them. Attenuating the connections between nodes is one form of control; keeping certain groups switched off—or completely unconnected to the network is also a form of control.

Soviet and American scientists were engaged in a deep rivalry to develop modern weapons.[20] In the 1960s, as the United States was developing ARPANET, Soviet scientists also proposed to unite the government's information management systems and create a universal information bank. The proposal came from Viktor Glushkov, who saw the project as a political as well as scientific reform to change decision making within the government. A network of computers would allow proposals and ideas from individuals to the government, not just better accounting of factual data for the state's economic planning work.[21]

Widespread access to information was seen as potentially undermining the Communist Party.[22] Recognizing the political implications of such a computer network, the Politburo scaled back Glushkov's proposal. While they approved network construction, they delayed its application to economic planning. Instead, each ministry built its own computer center and information system, which were mutually incompatible and therefore allowed each ministry to maintain control over its own information. This pattern continued until the 1990s when nongovernmental, private, commercial enterprises began innovating with Internet-type networks.[23]

Soviet computing was focused on defense programs—nuclear weapons, ballistic missiles, and anti-missile defense. It was government run for government objectives. There was no consideration of commercial uses and no scope for civilian applications.[24] The lack of a commercial market meant there were no devices to connect the incompatible computers in different parts for the government. There was no support for the hardware or software. Users had to write nearly all the programs themselves. In the 1970s, Soviet software had developed in isolated islands of information accessible only to technical experts.[25] By contrast, in the United States, starting in the 1950s, computers spread quickly from military to the business sector. Computers were not only used for mathematical calculations but also to process electronic data.[26] Moreover, American commercial vendors had an incentive to make computing power more easily accessible to a wider community, especially for business applications such as marketing, software, and customer service. Responding to customer concerns is one source of innovation.[27]

Rivalrous Soviet institutions hid their computer science findings from each other. Soviet scientists complained that it was easier to learn about foreign technologies than about other domestic computing.[28] By the early 1990s the government was deeply bifurcated between state-run computing and the rest of the sector, which was a mixture of private, foreign, and illegal activities.[29] Hindering connections between scientists not only blocked the growth of a trust community of researchers but also foreclosed the development of trust communities of civilians.

RUSSIA 1880: TSARS CENSOR PRINTING

The Soviet worries about the internal flows of information go back hundreds of years. In the seventeenth century, Peter the Great took control over printing from the Church. Book publishing went from around six titles a year in the 1690s to around fifty a year in the 1720s, from mostly religious to mostly government publications. In 1726, Catherine the Great installed a regime that required her prior approval for all new publications, and new titles dropped to about twenty a year.[30] However, Russia's intellectual life flourished with new journals, lodges, and schools, which created pressure to allow more publishing. In 1782, she started granting individuals the right to publish, ending the government's monopoly.[31]

In the nineteenth century, publisher Ivan Sytin made his fortune publishing educational books like Pushkin's Russian grammar in 1887 and Tolstoy's "Mediator" (*Posrednik*) series, a set of books based on the popular *lubki* morality tales targeting a bridge market between mass readers and the intelligentsia. Technically, Sytin's publishing house transitioned from books printed from woodcut blocks with hand-tinted images to lithography that enabled him to print images inexpensively. Pictures were essential to selling books.[32] While Sytin's work with Pushkin gained him prestige and he won awards for technical excellence in publishing, the Church opposed Tolstoy and banned nine of the Mediator tales.[33] Sytin expanded into newspapers. He bought the *Russian Word* and transformed it into the most widely read newspaper in Russia.[34] While *Word* was not subject to pre-publication censorship, it was under constant review and frequently resisted the Moscow press committee's discipline. In the early 1900s, when covering the Japan-Russia War, *Word* released its papers to the censorship committee and the public at the same time, precluding pre-publication censorship. In January 1905, when the newspaper complied with the government's reporting ban on Bloody Sunday, a workers' strike harshly suppressed by the tsarist regime, the newspaper's workers walked off the job in protest.[35] Another time, when Rasputin, fabled advisor to the tsar's family and contributor to its waning legitimacy, died on December 17, 1916, *Word* made it the headline story in spite of bans by the censors.[36] In 1917, the fiftieth anniversary year of Sytin's printing empire, the Bolsheviks won the Russian Revolution. While Sytin had supported the Russian war effort against Germany,[37] he and the *Word* had criticized the Bolsheviks. Lenin sequestered Sytin and nationalized his presses. The presses that once printed *Word* began printing *Isvestia,* one of the Soviet Communist Party's vanguard newspapers.[38]

The Russians and Soviets are not unique in their efforts to limit flows of information. In the twenty-first century, there are catalogs of governments that filter online content and their techniques; many countries have such policies for national security reasons.[39] Pakistan blocks websites of independence movements. South Korea blocks North Korean websites. China blocks sites of groups that oppose the government, including minorities and religions. Iran blocks online publications that challenge the government or its religion.[40] In authoritarian countries, at the first sign of a protest a common first reaction from the government is to shut access to the Internet and applications, such as happened in Burma/Myanmar in 2007 when monks across the country protested. In all these cases, people innovated with sharing information on the Internet. Governments then blocked the information, and hackers broke down government blocks. As news of the information wars spread, issues once confined to local trust communities became national and international in scope.

Information, Ideas, and Delivering Public Services

Governments not only block information but also promote certain kinds of information. For example, many governments distribute alerts about national disasters and other public emergencies as part of their public security policies. In many countries these are carried on television, radio, and possibly other media networks. Substantial coordination and cooperation is required between government, commercial, and other organizations. The impact of these networks is most felt when they fail, of course. Furthermore, the absence of such information distribution in times of public crisis is a sign of poor governance.

The effects of a tsunami that struck South and Southeast Asia in 2004 were discussed in chapter 2. Some governments had emergency communications systems that alerted them to the impending tsunami, others did not. Consequently, scientists in Hawaii and Japan, which were not heavily affected by tsunami waves, had the emergency information but were unable to connect quickly with their counterparts in Thailand, Indonesia, Sri Lanka, India, and other countries who were hit unprepared. This particular disaster highlighted a weakness in these countries' public safety systems, a responsibility that people expect their governments to fulfill. One of the salient criticisms against the U.S. government's response in the aftermath of Hurricane Katrina in 2005 was the failure to provide good, clear information to the public. In the following elections, citizens voted out old leaders and brought in new ones in part based on this failure. Another category of information is public health, where there is often a clear consensus that better dissemination of information and ideas can improve the overall welfare of society in a measurable and tangible way. This trust community of governments, ordinary citizens, and intermediary organizations is tied together by a patchwork of several communications technologies, nationally organized and globally coordinated.

Just as in the control of infrastructure, governments seeking to control information and ideas in the interest of national security may not succeed. Whether they succeed in large measure depends on the reaction of the people they govern. If there is substantial consensus that certain information, such as health warnings, should be distributed, or should be restricted, such as advance warning of battle plans, then the policies are more likely to be successful. Should there be substantial resistance, such as generalized ban on political speech, then it is likely that people will subvert the policies. Subverting policies may actually open up commercial opportunities for entrepreneurial rebels. As powerful as governments may be, they are still subject to the fundamental interactivity of communications networks. Even mass propaganda that is one-way from a technical perspective is two-way in terms of effectiveness—the audience decides for itself the meaning of the message, not the sender.

Much research focuses on how governments restrict information. However, quite commonly governments, both authoritarian and democratic, do promote and restrict certain kinds of information that promotes general welfare. A better understanding of how governments use information well could contribute to a better understanding of how networks of communities are best held together.

GLOBAL 1990: WORLD HEALTH ORGANIZATION LEARNS FROM CROWDS

E-government is about using technology to improve the delivery of government services to the public and meeting the rising expectations of tech-savvy citizens. For example, in 1994 the World Health Organization (WHO) was a minor player in helping India with a bubonic plague outbreak in Surat; whereas in 2009 it was at the center of an international campaign to combat the H1N1 flu virus. In the meantime, WHO shifted away from relying only on government-submitted information to taking into account all publicly available information. WHO also updated its use of information technology by employing tools like Global Public Health Intelligence Network (GPHIN), which uses a web crawler to collect Internet reports of unusual disease outbreaks, and ProMED, a moderated system for reporting infectious and chemical disease incidents. Both changes in information policy and technology tools undermine the nation states' control in the WHO but improve the effectiveness of the institution. Consequently, the WHO is a more visible and important player in international public health today than before.[41]

UNITED STATES 1960: INTERNAL REVENUE SERVICE GETS A COMPUTER

The public does not expect to love the tax service, but it does expect it to be competent. The U.S. International Revenue Service (IRS) reported that in 2012, 80 percent of individual tax returns were filed electronically.[42] This transformation began in the 1960s when the agencies first brought in computers for internal processing. In the 1980s a botched information technology transition, combined with inadequate capacity and staff, generated a crisis; 60 percent of returns were delayed. As the Internet grew in the 1990s, the IRS posted tax forms online and allowed electronic filing for businesses and tax professionals. In 1997, half of all tax returns in the United States were prepared on a computer, printed out, and then manually re-entered into IRS computers. In 2000 the government re-organized the IRS and accelerated the push to e-filing.[43] While the IRS was behind in the early 2000s, the growth in IRS e-filing has since kept up with the growth of Internet use (fig. 4.1).[44]

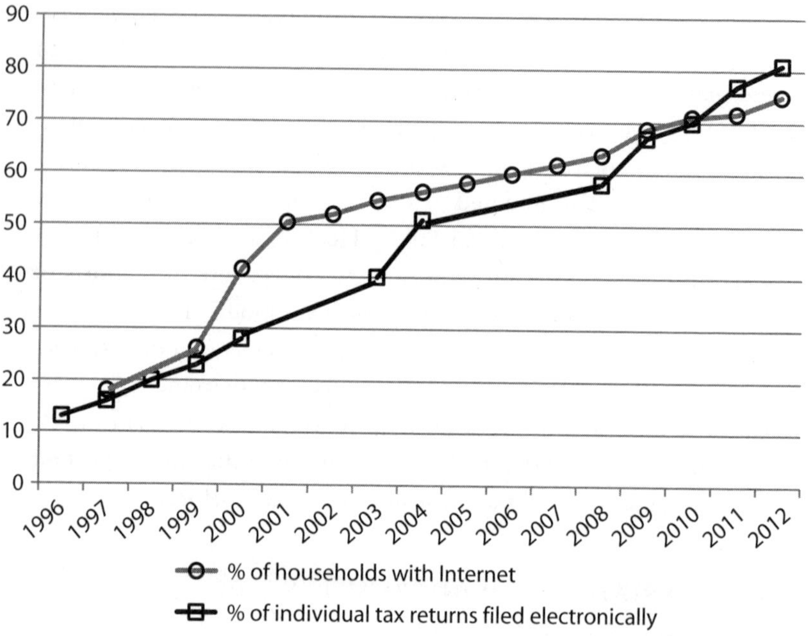

FIGURE 4.1. United States: E-filing and Internet subscribership. *Source:* International Telecommunications Union and U.S. Internal Revenue Service.

Both the WHO and the IRS follow typical stages in e-government development. First, the Internet is a billboard to post information for the public. Second, the public can search and access some data on the website. Third, the government delivers services online—issuing licenses and processing applications. Fourth, using the Internet to interact with and be accountable to the public.[45] They shifted from simply broadcasting information from the Internet to systems that allow them to take in substantial and varied information from a broader community than before.

Information, Ideas, and National Identity

INDIA 1987: DOORDARSHAN TV'S *RAMAYANA* RECASTS POLITICS

When *Ramayana,* a runaway hit television series in India, aired, "city streets and marketplaces were empty on Sunday mornings. Events advertised for Sundays were careful to mention: 'To be held after *Ramayana.*' Crowds gathered around every wayside television set, though few could have seen much on the small black and white sets with so many present. Engine drivers were reported to depart from their schedules, stopping their trains at stations en route if necessary, in order to watch."[46] The public broadcaster Doordarshan serialized *Ramayana,* one of the two major Hindu epics, in an effort to appeal to the Hindu middle class in the 1980s.[47] After *Ramayana*'s first episode aired in 1987, there were 78 weekly episodes in total, a blockbuster hit.[48] Market research companies estimated the audience at from 40 million to 80 million per week over a few months, bringing street traffic and markets to a standstill during its broadcast time. *Ramayana* was a runaway commercial success and a breakthrough for Doordarshan with high viewership across linguistically diverse regions.[49]

The state-sponsored television network participated in the transformation of identity politics, possibly inadvertently. Doordarshan, the public broadcaster in India, was established in the 1960s with the goal of creating an Indian national identity and promoting national integration. Originally, this was to be a secular identity, and programming emphasized national culture, a secular India united by English and Hindi languages, with mixed results. On the one hand, from the beginning Doordarshan's broadcasts intensified feelings of conflict among regions, between rural and urban communities, and between rich and poor.[50] On the other hand, sports have been very successful in unifying national consciousness. The broadcast in 1982 of the Ninth Asian Games made color televisions very popular, and Doordarshan viewership increased.[51]

The broadcast of the *Ramayana* serial coincided with the rise of the Hindu nationalist movement in India, a movement that has fed communal tension and violence. For example, the activists of the Birth of Ram (Ram Janmabhumi) appropriated visual images from television in their demonstrations. In one protest procession from Delhi to Ayodhya, participants dressed to look like the television versions of Ram and his brother Lakshman."[52] Processions like these, for example, culminated in December 1992 in the destruction of a mosque at Ayodhya by Hindu fanatics who wanted to build a temple at Ram's birthplace. These events led to communal riots in cities across India, resulting in about 2,000 dead and 7,000 injured.[53] These extremely popular programs like *Ramayana* provided the Hindu nationalist movement with powerful language, images, and visual symbols, essentially the cultural infrastructure, to distribute their ideas. By enabling the consolidation of Hindu nationalism, this television show contributed to reconfiguring the discourse of nation, culture, and community.[54] While the television serial did not cause the rise of this violence, the program gave the movement prestige, visual symbols, and a language to express itself. This Doordarshan production of *Ramayana* still resonates in India's politics today. Its great popularity fixed in the public mind a particular interpretation of the *Ramayana,* a legend with several versions and a variety of interpretations.[55]

Such iconic cultural imagery emerges not only from government programming to politics but also from popular genres as well. Hit movies with historical themes often supplant the actual historical facts in the public mind. Television programs tied together a trust community launched by the public broadcaster but given meaning by ordinary viewers and activists looking for a repertoire of images to transform the national political scene, sometimes with implications for political dialogue for years to come.

Commercial advertising jingles can pass into political dialogue. For example, nearly coincident with Doordarshan's *Ramayana* broadcast in India, a commercial for U.S. hamburger restaurant chain Wendy's popularized the question, "Where's the beef?" suggesting that its competitors burgers had less beef. The slogan leapt to the U.S. presidential campaign of 1984 where one candidate, Walter Mondale, asked another, "Where's the beef?" suggesting his opponent's politics lacked substance.[56] These memes, ideas laden with cultural and symbolic significance, have a life of their own and cannot be controlled by their creators. The meme's development follows a trajectory that interacts with multiple interpretations by different groups. Thus, any effort by governments to launch and push memes can have unintended consequences.

Governments frequently use information and ideas delivered over communications networks to create and promote a national identity. The most prominent examples are public broadcasters, which range from news reporting scrupulously protected from political interference to networks designed to deliver only the government perspective.[57] In all instances, a clear measure of success is whether anyone is watching and listening. Audience engagement is a sign that the government has successfully nurtured a trust community. As news and entertainment programming moves to the Internet and is received on computers and mobile phones, these public broadcasters must adapt in order to continue engaging their audience, converting them from viewers and listeners to participants in an interactive online world.

UK 1938: BBC WORLD SERVICE RADIO EMBODIES EMPIRE

Internationally, the public broadcaster most successful in engaging its audience is the British Broadcasting Corporation (BBC). Originally established in the era of the British Empire, the BBC successfully framed the worldview of its listeners throughout the colonies. Integral to its success was the accuracy of its reporting; its integrity won the trust of audiences scattered throughout the globe. Simultaneously, the BBC serves as an image-maker of Great Britain to the rest of the world. Outside Great Britain, the BBC is seen as an authoritative, reliable source of news and not the propaganda tool of the British government and has been highly regarded by audiences in regions around the world. According to the BBC, half of affluent adults in Africa watch the BBC at least once a month, a higher percentage than in other regions outside Great Britain.[58] In English-speaking countries on the continent, the BBC is the "voice of record."

Similarly, in the Middle East, the BBC Arabic service became important in light of the propagandistic nature of available news services sponsored by other national governments. The Arabic radio service began in 1938. In its inaugural broadcast, the editor included a report about a Palestinian who was hanged for keeping weapons in his home. At the time, opposing British policy for establishing a Jewish country in Palestine was a crime. This reminder of the harsh edge of their policy embarrassed British who had invited Arab friends to hear the BBC's first Arabic broadcast. However, the BBC included the story as a marker that it intended to produce an honest and factual service, not be the press office of the British government.[59] While maintaining substantial autonomy from the British Foreign Office's influence, the BBC's goal to keep its reporting fair and balanced was tested over the years, from the Suez crisis to the Iraq war.[60]

It is at home where the BBC's tension between its government ownership and its obligation to provide an independent news service is greatest. From its inception, the government sought to keep controversial material off the BBC. Sometimes leaders in the ruling party successfully prevented the speeches of opposition party politicians from airing over the service. While over time the BBC established independence from direct political interference, there have been crises. For example, during the Cold War in 1965, a documentary called War Game that illustrated the likely effects of a nuclear war was determined by the British government to be too frightening to air and too pointedly demonstrating the weaknesses of the civil defense programs designed to protect people from the effects of nuclear war.

More recently, the British government's Hutton Inquiry (2003–4) investigated the death of David Kelly, a technical expert on issues central to the Iraq war. The BBC claimed he was the source for their reporting that Prime Minister Tony Blair had exaggerated the danger that Iraq posed prior to the United States and Great Britain's invasion of Iraq in 2003. As publicity mounted, Kelly committed suicide. The investigation of the BBC's relationship with Kelly is a reminder that even while the BBC maintains its reputation for integrity and fairness, its status is in continual negotiation.[61] BBC news, which travels over radio, television, and now the Internet, ties together a national and international audience community and also is a key facet of the British national identity.

QATAR 1996: AL-JAZEERA SATELLITE TV RAISES QATAR'S PROFILE

A newer public broadcaster is Qatar's Al-Jazeera. Qatar's launch of the Al-Jazeera news network remade the country's image in the Middle East region and in the world.[62] The Middle East lacked an Arabic news service from the region that provided news without strong political influence from the broadcaster's home country. This vacuum was filled by Al-Jazeera.

Qatar's government launched Al-Jazeera in 1996 as part of a larger effort to raise Qatar's international profile in the world. In the 1990s, its leaders took action to distinguish Qatar from other Gulf States. Qatar attracted leading international universities to establish campuses in Doha and hosted international events ranging from economic meetings like the World Trade Organization to sports events like the Asian Games. Al-Jazeera was established in this context of Qatar's unique foreign policy strategy.

In 1995, the emir lifted press censorship, and in 1996 Al-Jazeera news service was launched with his funding. Al-Jazeera revolutionized the news in the Middle East. For the first time, television audiences have access to professional, in-depth coverage of political issues, with more viewpoints than just the government's, presented in exciting, attractive formats. When many state-run news agencies waited for government instruction before broadcasting, Al-Jazeera introduced real time news coverage. For example, it was the only Arabic network broadcasting from Baghdad when the United States and allied forces launched their attack in 2003. It was the first non-Palestinian Arab network to air interviews with Israeli officials. Al-Jazeera provided facts, which meant competing news agencies could not get away without reporting them. Also, Al-Jazeera's talk show formats, especially *The Opposite Direction,* broke taboos—now these talk shows are a staple of the news scene.[63]

Al-Jazeera's news coverage does face constraints. However, as in the case of the BBC, the constraints are primarily at home. Al-Jazeera does not report on the power struggles between its largest investor, Sheikh Hamad, and his father or on the pace of democratic electoral reform in Qatar, for example.[64] Also, Al-Jazeera can be seen as furthering Qatar's foreign policy objectives by challenging Saudi Arabia authority through interviews and news stories.[65] Al-Jazeera news distributed online and over television ties together a national, regional, and increasingly international audience community. It has raised the profile of Qatar in Middle Eastern and global politics. In 2011 as revolutions broke out across Tunisia, Egypt, Bahrain, and Syria, Al-Jazeera collected videos and news posted to social media by the public, especially where its reporters were banned, and powerfully redistributed the information to a transnational audience that it had developed over the previous fifteen years.[66] It became an important information broker, essential for those without Internet access.

DISCUSSION: BBC, DOORDARSHAN, AND AL-JAZEERA

Public broadcasters work in a market that is commercially and technologically evolving. For governments, there are both financial costs and benefits. Doordarshan's departure from a secular programming with *Ramayana* gave it a major commercial success, which it then replicated with similar serials. At the time, Doordarshan was facing competition from satellite programming from abroad. The BBC was a unique venture of its sort when it launched. Now it faces competition from other international news networks, most of which are commercial. BBC's international service generates important revenue; however, most of the BBC's income is from taxes on the British people. To the extent that the BBC benefits millions of people outside of Great Britain, how long will the British public be willing to sustain funding?[67] Al-Jazeera is yet to be commercially viable and still relies on investment from Qatar's leader. Satellite programming in the Middle East is still in its early stages, and the final commercial direction is unclear. Al-Jazeera is developing its sport broadcasting franchise, a possibly lucrative niche yet to be filled in Middle East broadcasting.[68]

Conclusion

Historically, governments have used communications infrastructure to deliver ideas and information that hold together the nation. Government institutions have specific goals, formal rules and enforcement capabilities, and a worldview that holds together its members. National identity, economic development, and national security are the typical goals of communications policy. The scope of the infrastructure over which the ideas and information travel is determined by whether it was constructed to meet market demands or political goals. The degree to which the ideas and information are well received by citizens also depends on whether it engages their interests rather than reflecting the government's views. The strength of the trust community led by the government depends on its ability to successfully foster infrastructure and enable ideas that actively engage citizens.

National security is one major goal of governments' communications policies. Militaries are often deeply involved in communications infrastructure. In the United States, the telegraph was far more commercially developed in the North, and during the Civil War it played a key role in supporting the Union forces led by Abraham Lincoln against the Confederate forces. Lincoln and his generals made full use of the telegraph to coordinate strategy and tactics. In Brazil, the telegraph was constructed by the military, especially the sections that spread into the Amazon. However, without a commercial market to support the infrastructure, it was poorly maintained and eventually collapsed. Beyond infrastructure, information policy is also a national security concern. In the twenty-first century, information ministries in countries like China, Iran, Pakistan, and Korea block Internet websites in order to prevent certain ideas from coming into the country and undermining the regime. However, just the reverse is also true: governments are improving technology to provide basic services, from better warnings of natural disasters to collecting taxes.

Economic development is a second major goal of governments' communications policies. The largest national telecommunications network in the world is in China. Economic reform policies drove construction in the 1980s. Even for authoritarian regimes, economic benefits from better communication can override the risk that outside information may undermine the regime. In the United States, while military research and investment started the development of the Internet in the 1960s, it was the commercialization of the network—the government letting go of control—that allowed the Internet to become the global communications infrastructure of the twenty-first century.

Fostering national identity is a third major goal of governments' communications policies. Canada's nationwide radio broadcast for its Diamond Jubilee in 1927 tied together the people of its eastern and western coasts, underscoring its transcontinental character. Public broadcasters such as the British Broadcasting Corporation (BBC), Al-Jazeera for Qatar, and Doordarshan for India, use their programming to create narratives for the nation and for presenting an image of the country to the rest of the world. With these ideas and information, the public broadcasters must not simply voice the views of the government but instead engage audiences. They must find a partner in their viewers and listeners, not just subjects.

Trust through interaction builds successful infrastructure and information networks. In China the telecom infrastructure is built to meet millions of users' demands; therefore, it is robust enough to deliver the government's information and ideas. If the Internet had few users, it would be useless as a tool to survey the public mood. Al-Jazeera's programming engages audiences throughout the region. Therefore, Qatar has the opportunity to influence Middle East politics. People watch Al-Jazeera because it presents facts that other broadcasters have skipped, and its talk shows give voice to the previously voiceless. For Qatar, Al-Jazeera and its audience are a trust community, a source of political capital and ammunition worth as much as money or military might.

In contrast with activists who struggle to build the resources, tie together a network, and create a community, states consist of institutions with formal rules, memberships, defined scope of action, and enforcement mechanisms. Their challenge is to maintain these institutions' trust communities and renew their network of participants by refreshing its fund of social capital, trust, and sense of identity. Success depends on whether governments engage people to participate in its network. This applies to both infrastructure and information. The telegraph thrives only if people send messages. A television program only has value if people watch it. Governments build communications infrastructure, disseminate information, and build trust communities. With these trust communities, they are powerful. Without them, governments risk irrelevance and illegitimacy, and other trust communities will compete and challenge them.

PART THREE

Trust Communities in Politics

CHAPTER 5

Technology + Trust = Political Influence

IN THE 1990S when the Internet first went commercial, people speculated that the flood of information and ideas would naturally, in time, bring down dictatorships, undermine national borders, and homogenize cultures. Twenty years later dictatorships persist, national governments are strong, and local cultures flourish. Yet the Internet has deeply affected politics. How, precisely? In terms of politics, the Internet changes the constellation of people we trust. We can interact with more people, more easily, with greater regularity. We can experiment with joining new communities. When one community fits, we trust, we join, and we add a new layer to our identity. These trust communities can be powerful political actors. More research on how people use technology to build trust will enrich our understanding of politics. Based on the case studies in this book, there are several findings that lay the ground for future work.

First, technology enables an individual to connect with more people in a practical, everyday way. This creates the opportunity over time for positive, reciprocal activity among these newly connected people. This work builds trust. Once individuals trust each other, there is the opportunity to build social capital, the kind of intangible resource that makes collective action possible.

Trust communities are one in a continuum of social organizations. At one end are networks, which are largely nonhierarchical groups of loosely connected individuals. At the other end are institutions with hierarchies, formal rules and procedures, membership lists, budgets, and mission statements. In between are trust communities—groups with a shared cause or identity, held together by technology rather than common locale. They may have leaders but not necessarily a strong hierarchy.

For institutions, technology also opens the door to more trust communities. Technology enables them to interact differently with their members and to expand to new ones. Emerging trust communities may compete with established ones. For example, governments invest in telephone, radio, and television systems that physically and ideologically support the imagined community of a "nation." New technologies challenge old trust communities, requiring governments to adapt if their institutions are to remain relevant.

Second, popular technologies, not niche or elite tools, are more likely to have political impact. Once people trust a technology for shopping or entertainment, they may escalate to using it for politics. Therefore, the economic incentives that drive commercial development of a technology have an inextricable impact on the politics of the age.

Third, trust builds in a community when members repeatedly interact and reciprocate. An ad-sponsored television talk show has a studio audience, a viewing audience, and a commercial clientele to cater to. Here there is a lot of interactivity. This is a trust community. Online, similar interactivity is possible at more individualized level and with a more diverse crowd. A state-run propaganda program is the opposite. While in principle the viewing audience may be large, the public may ignore its message or—even worse—reinterpret it to mean the opposite. No trust community here.

Fourth, trust communities often have diverse memberships by conventional measures like class, ethnicity, or religion. Ideas and information hold together trust communities; it is often easy to opt in and out.

Fifth, while trust communities are bound by technology, technology is not their currency. What members exchange is information; what they share are ideas. For them, information and ideas are their capital and their ammunition. This is the contrast between trust communities and other analytical lenses like economic class, religion, or ethnicity. Understanding information and ideas as a basis of power sheds light on amorphous ideas like soft power and reveals ways to measure it, similar to aspects of military and economic might.

Sixth, trust communities are potential units of political power. In addition to using categories like class, ethnicity, and religion to dissect society into understandable constituencies, it is possible to use trust communities as an analytical lens. Communities connected by a particular newspaper or television show bear certain characteristics; communities connected by an online social media application bear certain traits. Within these trust communities, if there is enough trust, social capital may build, and if so, collective action may be possible. A nation is one of the many trust communities competing for an individual's commitment. While the nation is still a powerful idea, there are other trust communities to rival it.

Trust Communities—Opportunities for Individuals and Institutions

Reciprocity builds trust, and technology enables reciprocity among a wider array of people. Ostrom and Walker (2003) demonstrate through their empirical experiments that communication substantially increases cooperation in many types of social dilemmas. Their experimental work demonstrates that in the absence of an external enforcer, people cooperate when there is opportunity for exchange, especially if the partner can reciprocate. If participants communicate more frequently, they are more likely to cooperate. Communication transfers knowledge about the optimal strategy, allows exchange of promises, increases mutual trust and expectations, increases the participant's personal payoff, reinforces prior normative rules, and facilitates development of a group identity.[1] When people have a history of past cooperation, they expect cooperation in the future.[2]

Many major works on collective action have reviewed the possibility of genuine collective action primarily organized over the Internet and concluded that it is not possible. In 2003 McAdam argued that collective action is most likely to take place within organizations that are already stable—like civic groups and neighborhoods.[3] Also in 2003, Putnam, Feldstein, and Cohen emphasized face-to-face communications in small settings. Their study of the Internet service Craigslist suggested that it was unusual for online communities to build social capital.[4] In a 2003 publication, Ostrom and Walker expected that whether communication is face-to-face interaction or mediated should make no difference. Instead, their experiments showed that face-to-face communication increases cooperation more than does computerized communications. Computerized communication did not achieve same level of efficiency.[5]

Putnam and his colleagues demonstrate that one of the most powerful ways for people to share attitudes, values, and identities is through stories.[6] People like to tell their own stories and listen to others as well. Personal narratives create empathy and allow people to discover commonalities. These narratives express needs and build understanding. Sharing stories is a powerful recipe for building social capital and social motivation.

This may reconcile the skepticism these scholars of cooperation have with the experience of Internet users. When Ostrom and Walker ran their experiments, the computer communication they tested was simple text on screen. With the Internet, however, computerized communication is much richer than before. Now computer communication includes video communication that simulates face-to-face communication and allows a person to convey a thick self-portrait to others.

Today social media tools allow individuals to tell their personal narratives in newly complex ways. The simplest narrative includes a picture, a list of jobs, and membership in groups and associations. The more complex narratives reveal the individuals' friends and family, snapshots of life events across time, commentary from the individual's friends. Such a personal narrative, which can be perused in a quarter of an hour online, might take several hours to convey in a simple face-to-face conversation. So while face-to-face communications are unique, they are no longer the only way to have the multi-stranded communication necessary to build trust and cooperation.[7] In the 2011 popular uprisings in Egypt and Tunisia, social media allowed individuals to share with each other and broadcast to the public thick portraits of themselves and their political views.

Joining a trust community gives a person one more place to belong. Psychologists like Tom Tyler show that people want to create and maintain a favorable sense of themselves. They want to be members of groups they can take pride in, groups that are respected. They have an emotional need to belong.[8] For example, in the *TsunamiHelp* group volunteers wanted to belong to a group that actively helped victims. Many said they were not satisfied only contributing money to charitable organizations, especially since many doubted their financial integrity. The Internet enabled people with the common values and attitudes to come together. Once formed, the *TsunamiHelp* community became part of the volunteers' identity. The trust community adds a layer to a person's identity.

The trust community is a social space to see and be seen. Elisabeth Noelle-Neumann talks about "spirals of silence": "Wearing a campaign button, putting a bumper sticker on the car—these are ways of talking; not doing things, even if one has firm convictions, is a way of keeping quiet. Openly carrying around a newspaper which has a well-known political slant is a way of talking; keeping it out of sight, in a briefcase or beneath a less partisan paper, is a way of keeping it quiet."[9] Individuals have regard for the views of the public and the views of their community because they fear isolation, disrespect, and unpopularity.[10] Those who talk, encourage others to talk; those who are quiet, encourage others also to swallow their voices. For example, in the case of Taiwan, when cable television news openly discussed opposition views, it broke a spiral of silence on such discussions, an essential part of Taiwan's transition from authoritarian to democratic systems.

Emerging trust communities compete with established institutions for people's time and attention. Al-Jazeera trumped other state-run news agencies in the Middle East. The International Campaign to Ban Landmines created a more authoritative reporting service than governments. *TsunamiHelp* attracted more traffic than many international humanitarian websites. In all these cases, new trust communities with better ideas and information challenged established institutions.

Just being connected to other people does not automatically build social capital and lead to collective action. Just adopting new technology in government institutions does not guarantee greater engagement or support by the people for the state. It is no wonder that studies that seek simply to link the number of Internet users in a country to improved participatory politics have found no direct causality. Connection by technology alone does not transform politics; people must use that connection to learn about each other, to build understanding and trust, in order to create social capital and, finally, create the conditions for possible cooperation. Once a network of people connects, they become a community if they have built the trust to believe in a common identity and work toward a common cause. If that community acts together repeatedly, drawing on a set of practiced tools and techniques, then it is on its way to becoming an institution. For institutions to remain relevant, they must constantly renew their trust communities by maintaining trust and a sense of common cause among their members.

The Role of Capitalism

Most of the cases in this book involve a technology that was a great commercial success. In nineteenth-century China, activist Jing Yuanshan was also the business leader responsible for making a commercial success out of the Shanghai telegraph network, while the imperial court's telegraph system languished. Jing's telegram service had better office hours and cheaper prices. In the Philippines, protestors took advantage of two business developments to oust President Estrada in 2001. In the 1990s, mobile phone company Globe Telecom introduced cheap texting (short message service or SMS) as a strategy to gain market share against its competitors. This triggered a texting price war and made the Philippines a world leader in texting in terms of traffic volume.[11] Second, while the corruption saga was unfolding, tabloid newspapers sold huge numbers if they were willing to risk covering President Estrada's scandalous personal life. In Taiwan, cable television entrepreneurs invested in infrastructure in order to make money selling entertainment programming. Carrying news programs that reflected their opposition politics was a tandem activity.

Business and politics are intertwined. Benedict Anderson argued that in the early twentieth century print capitalism—especially the business of newspapers and popular novels—contributed to spreading the idea of a nation. Both created for their readers a sense of simultaneity. A daily newspaper contains what I need to know to chat with friends. Similarly, the novel contains the characters and stories that are hot topics in my community. Note that the *popular* newspaper and the *popular* novel create this sense of community, connecting capitalism to the flourishing of the nation.[12] Anderson's work links personal identity and the world of commerce. Today the equivalent to the newspaper is electronic news—television, radio, and Internet. The novel's equivalent is the movie, the television drama, comedy, and how-to show. Part of our identity draws on what we choose to read and watch.

We live in a world of media capitalism, just a step beyond Anderson's print capitalism.[13] Parallel to the stories about email, social media, and other Internet services on politics are narratives of technical inventions, investors, business strategies, and ties among companies, civil society, and government. Furthermore, popular media should have a privileged place in research seeking to understand the political impact of technology. Popular media recognize the preeminence of the consumer. The more energetic the quest for popularity, the greater the importance of the audience—the reader, the viewer, the listener—the kind of engagement required for the development of a trust community.

Engagement, Participation, and Interactivity

In a network people are loosely connected, but in a trust community people engage because they have a common identity or a common cause. Engagement means people interact with each other; there is no central point that directs a passive and accepting audience. Instead, there are many points; there are varying messages; the audience is active and may accept, reinterpret, or resist information and ideas.

In her landmark study of the American television series *Dallas,* which was an international broadcast phenomenon, Ien Ang demonstrated that the recipient interprets meaning, not the producer of programming. This long-running American soap opera about a wealthy Texas oil clan was hugely popular around the world. However, people in different societies had contrasting interpretations—was it about the showy glamour of oil wealth, was it the unbelievable family conflicts, or was it the big Texas culture? It all depended on the viewers, not the producers of the television program.[14]

Technology facilitates more interactivity. With newspapers, a reader writes a letter to the editor and hopes for a response or the slimmer possibility of getting printed in the next day's or next week's edition. For television shows, people call in to complain or join a television show audience, but there are constraints on airtime, seats in the audience, and the patience of the producers. Online media greatly reduce the constraints on interactive participation. It is far easier and quicker for audiences to react to a news report by commenting on a website; often audience members can comment on each other's comments.

From a different perspective, Florence Passy, in her study of social networks for political purposes, echoes Ang's same conclusion that members' engagement creates the meaning of the community.[15] In their study of activists in a Swiss movement that advocated development aid, Passy and Marco Giugni examine why individuals join a group and decide how intensely to engage. First, there is the environment. People around them talk about the issue, define it, establish values and priorities. This is a long-term process. Then, they connect to the movement through social networks. This can be quick. The participants must weigh how much to get involved—the risks, the rewards, the legitimacy of the cause, and their available time. Giugni and Passy found that individuals change the intensity of their engagement over time, which in turn shapes the community itself. The trust community influences the participants, but the participants also influence the contours of the community.[16]

This interaction among producers of ideas and receivers of ideas illuminates the international relations notion of "soft power," a country's ability to achieve goals through persuasion by virtue of its values, practices, and culture.[17] The source of soft power is not only in the institutions projecting power but also in the people that are subject to its influence. The public broadcasters BBC and Al-Jazeera both demonstrate that their state sponsors gain in influence by providing a news service that focuses not on merely relaying the views of their owners but on delivering to audiences the information and ideas they want. The influence of the BBC and Al-Jazeera depends on their ability to serve their customers, not on the strength of their state sponsors. Yet when they succeed in serving their customers, the benefits redound to the state sponsor.

Trust Communities and Diversity

Trust communities often appear heterogeneous by the typical standards used to analyze social groups. Members may differ in geographic location, socioeconomic class, ethnicity, nationality, or religion, for example. Since people choose to engage in a trust community, the tie that binds them together is a common cause or identity that they choose for themselves, not one defined for them.

The examples in this book bear out this heterogeneity among trust communities. In the Al-Jazeera case, the trust community members belong to different socioeconomic classes, religions, nationalities, and languages. They are bound by their interest in the Al-Jazeera point-of-view. In the *TsunamiHelp* case, volunteers from the Great Britain, Sri Lanka, United States, Australia, Thailand, and other places quickly join volunteers in India. They come together online to collect humanitarian aid information for and about victims of a large tsunami which hit South and Southeast Asia in 2004. The volunteers were from different countries, different professions, and with different attitudes toward volunteering. In the case of the International Campaign to Ban Landmines, the volunteers were not only individuals but also organizations. A heterogeneous group of human rights organizations, relief agencies, aid organizations for landmine victims, governments, and military were all part of the trust community of this campaign.

Several researchers underscore the benefits to society from heterogeneous networks. In her work Kathryn Sikkink recognizes this heterogeneity as characteristic of networks as a form of social organization. She thinks of networks as consisting of a variety of kinds of members—individuals, civil society groups, and states. She also emphasizes that information and learning tie together the network and give it meaning. The voluntary nature of these networks is an essential characteristic. Members can always exit.[18] The easier it is for members to exit a trust community, the harder it has to work to keep them participating.

Early studies of online activism often highlight that members were so heterogeneous that their change agenda was unclear, thus leading to no outcomes other than publicity for the movement. This was particularly salient in the global social justice movements that accompanied international meetings and World Trade Organization meetings in the early 2000s. Anarchists were linked with environmentalists, labor activists, and other civil society groups. The connections were loose; in some cases more like a network than a coherent community.[19]

The weakness of a trust community may be its diversity, but its strength is in its members' freewill commitment to an idea. Putnam's work on social capital underscores the difference between two kinds of social capital—bridging and bonding. Bonding is among like people; bridging is among different people. Putnam believes bridging is more difficult to develop but more important for the good functioning of society.[20] Trust communities with diverse memberships could build bridging social capital. While trust communities may be heterogeneous in terms of the usual social labels of ethnicity, politics, or class, for example, they are homogeneous in their commitment to the common cause of the group and its tools and techniques to take action.

Information and Ideas as a Source of Power

For trust communities, ideas and information are as politically important as capital, as land and natural resources, as military might. To build its identity as a nation, Canadians constructed a telephone network stretching east-west across Canada to resist the north-south pull from the United States. The BBC news service today is still a forceful representation of Great Britain, surviving the collapse of the British Empire, its reputation for integrity intact. To mitigate the power of trust communities, the Soviet Union attempted to keep information segregated in its scientific communities in the 1960s by not linking their computer centers. The Qatar government's launch of the Al-Jazeera news network was just the opposite action. By supporting an information network with more and better reporting than other Arab state-sponsored networks, Qatar not only changes Middle East and global politics, but also enhances its status.

This idea of power as ideas and communication is not new. A leadership secure in its relationship to a trust community, comprised of voluntary participants held together by common cause or identity, can be extremely influential. Money and munitions are not required; the volunteers will contribute their own.

Manuel Castells argues that in a network society information has preeminent influence over other forces. The important power battles are over cultural matters, over the manipulation of the media, and fought with symbols and ideas.[21] From a political economy perspective, Yochai Benkler argues that one of the most important effect of networks is the decentralization of information and cultural industries. Big business no longer controls views and ideas of individuals and therefore power is redistributed.[22]

Sandra Braman argues that modern states are distinctive in their use of information as power. She views it as the fourth power. The others are instrumental power, such as military weapons and economic sanctions; structural power, like rulemaking for trade and environment; and symbolic power, such as democracy and freedom. The new power is informational power: controlling information, defining what people understand about the world, delimiting their vision, and narrowing their options.[23]

Charles Tilly discusses capital, coercion, commitment—three types of social forces that define cities and states. Coercion involves use of force; capital involves production and distribution of goods and services; commitment involves trust networks—people to call on for help, to rely on over the long run or over long distances. The more a government relies on commitment to rule, the better the life of the community. Capital also benefits from trust networks and commitment. Tilly notes the example of ethnic networks that trade over long distances.[24] Trust communities do not necessarily have capital or coercive power, but they do have commitment. Kathryn Sikkink notes that the power of the network is not in its structure or material resources but rather in its purpose. The power of the network resides in the power of the idea it projects onto the world.[25]

Trust Community as an Analytical Lens

Using trust communities to analyze political power connects the everyday life of ordinary citizens to national and international politics. Just as a person's job, income earned, and money spent connects them to the nation's Gross Domestic Product and to the international trade system, so also a person's trust communities connect their sense of themselves to national identity, political institutions, and ideological battles played out on the international stage.

The concept of "trust community" draws from a history of ideas that shows how countries are in part defined by their communications and media systems. In the 1960s, Karl Deutsch focused on the infrastructure aspect of *trust community,* drawing a physical boundary around the distribution of information and ideas. Deutsch defines communities as groups of people who communicate with each other and understand each other beyond "the mere interchange of goods and services." How well people communicate with each other reveals how well integrated the community is and whether it is becoming a nation.[26] Canada's transcontinental telephone network defined it as a nation, a community tied together East to West.

In the 1990s, Benedict Anderson's *Imagined Communities* studied the distribution of printed newspapers and books, defining the nation in part by the scope of these communication networks. But he takes the concept further. Not only can an idea hold together a community, but a narrative can hold together a nation. Anderson's idea of a nation is that it is an imagined political community, limited and sovereign. He defines it as imagined because even in small nations no citizen ever knows all the other citizens—these are connections made in the world of the mind, not in the world of the streets and buildings. It is limited in reach because no nation is thought of as encompassing the globe; there is always an "other," a foreigner, and outsider to the community. It is sovereign because this is the basis for government authority, not divine right or lineage.[27]

In the 2000s, Manfred Steger talks about a "global imaginary." The "imaginary" is the tapestry of shared ideas and outlooks that make it possible for people within a community to fit together, how things work within the community, the expectations individuals have for each other, the common symbols and narratives that tie them together. The BBC serves as the nation's pool of commonly accepted facts about the world, sunk deep into the identity of the British people, and as the foundation for a shared outlook. The BBC is so central because of the information and ideas it brings—not whether it is a radio, television, or Internet service. This tapestry of shared ideas and outlooks explains why national holidays have certain meanings and are celebrated with particular rituals. It explains why in times of crisis, a community frequently coalesces around a selection of possible actions modeled on actions taken before.[28]

The idea of a *trust community* brings together the imagined community of Steger and Anderson, and the tangible physical infrastructures of Deutsch's networks. The nation is one of the most powerful trust communities; it is the community that many people most closely identify, defend, and are willing to die for. National narratives run deep, creating a shared history for people who have never met each other. National purposes give common cause, a shared way of life worth defending against others.

While nations are still the preeminent trust community, they have never been the only ones. There are also the local trust communities we belong to—our towns, our counties, our sports associations, civic societies, and cultural appreciation clubs, which are a powerful draw on our time, attention, and loyalty. Also, there are global trust communities—now more easily accessible to all of us than before. Not just the local athletic league, but the World Cup and the Olympics are available to us. Not just the local theatrical society, but films and movies from around the world are reachable in a few clicks. Not only can we work on the city budget and local crime, but we can work on problems of world hunger and global justice by expressing our views, giving money, or contributing information. The growing number of options gives rise to the sense of growing globalization. However, the pull of the national trust community is still strong. If it adapts, it can prevail, but it now competes for time, attention, and loyalty from other trust communities.

Future Research

Most of the examples in this book are historical, and researching them involved teasing out the role of communications service and technology from accounts and analysis that were primarily focused on other aspects. However, for studies of recent and future trust communities, technology lends itself to research designs that marshal qualitative and quantitative evidence. Chapter 2, examining the *TsunamiHelp* episode, is an example of this kind of work, although doing research on that case four years after it took place was long enough for much empirical evidence to evaporate.

First, one line of inquiry is examining the impact of technology on people's ability to build trusting, reciprocal relationships with new networks of people, and the extent to which that builds social capital. The findings of social science experiments show that repeated interactions can generate at least reciprocity and possibly to trust. These, combined with psychologists' arguments that individuals build trust when they share each other's stories, provide clues on how to improve the study of activist groups. Research on collective action can look at the frequency, character, and intensity of interactions among members of the community. These interactions are face-to-face and mediated. Among the mediated interactions, those taking place online can be measured. In the case of social media, the interactions can also be characterized as primarily transactional—let's meet in the public square at noon on Tuesday—or sharing narrative—this is the story of my life. These narratives can be further characterized into categories such as narratives that tell a person's individual background, narratives that explain how the world works, narratives that explain how to get certain things done, and other categories that give a sense of the type of community that is being created.

Second, another line of inquiry is to examine the degree to which trust communities with a repertoire of techniques and tools, turn into institutions with clear hierarchies, rules, and procedures for taking action. Asking a trust community's members about their history prior to coming together and examining how often over time their collective actions have taken place provides a useful perspective on whether an organization is likely to become more like a formal institution.

Third, for existing institutions the reverse of the first two steps can be a useful examination. Is the institution adapting to new or modified trust communities? Are their tools and techniques, goals and styles of participation also changing?

Fourth, the influence of commerce and technology and politics should not be overlooked. For example, most research divides the study of popular culture and communications from serious news, political discussions, and debates. However, the two overlap. Not only is there tension in infrastructure policy where both governments and private sector invest and operate services, but similarly, there are tensions in content policy. With greater ability to measure audience participation, as it is usually termed in entertainment circles, or citizen engagement, as it is usually described in political studies, the relative reach of government content can be measured against entertainment content. A related question is whether tools and techniques learned in one arena—voting for the best singer in television contest—can be transferred to another—voting for the best candidate in a city mayoral election.

Fifth, in investigating trust communities, whether it be a hactivist group or a government public broadcaster, researchers should study not only the message but also the response. With the Internet and its new applications, this is easier than before. With earlier technologies, the primary tool to study audience response would be to survey them—time consuming and expensive in the best of circumstances and nearly impossible in historical cases. Now, in addition to surveys, researchers can measure traffic in frequency and volume. Furthermore, many new media applications allow audiences to articulate responses directly through simple votes, ratings, or comments.

Sixth, the membership of a trust community should be analyzed for commonalities and dissimilarities. Initial studies of online trust communities suggest that membership is more heterogeneous and less hierarchical than in communities largely generated in face-to-face meetings. Investigating empirically whether this is systematically the case would contribute to understanding the degree to which Internet trust communities differ from pre-Internet trust communities. If tags like socioeconomic class, religion, and ethnicity are not homogeneous across the trust community, what categories are? In the *TsunamiHelp* case, it is fair to say that all the volunteers had a high level of comfort with complex computer and Internet skills. This also presupposes a high level of education. However, otherwise they ranged in age, ethnicity, profession, and religion. This book hypothesizes that the heterogeneity results from the primacy of ideas and information in drawing together members of online trust communities, but is there some other commonality?

Conclusion

To understand politics, we divide people into different groups—by income, by ethnicity, by profession, by religion, by nationality—and explain people's behavior according to the interests their groups have. Trust communities are another way to group people together and understand how they act politically. Trust communities are groups of people who share a common cause, a common interest, or common identity and are physically connected with each other by a communications infrastructure. This can be a group of volunteers around the world sharing information over the Internet, a nation of people following the evening television newscast, or a bunch of neighbors keeping up to date with a community newspaper.

Membership in a trust community is voluntary, and participants are heterogeneous by the standards typically used to understand politics. Viewers of an evening newscast may be different in terms of income and religion, for example. Volunteers organized over the Internet may be of different nationalities and political parties. The neighbors may be of different professions and ethnicities. However, because they have a shared cause and purpose they belong to the same trust community. The evening news community seeks to understand the current events of the day; the volunteers may be joined in preventing flu epidemics around the world; the neighbors may be organized to improve traffic safety. Beyond the obvious members of the trust community, there are others who are included. The governments that the volunteers are trying to convince are part of the trust community; they are on the receiving end of the volunteers' communications, emails, and television interviews. The reporters and producers of the evening newscast are part of the trust community; they are an integral part of the common goal of understanding current events. The police and local officials responsible for traffic safety are also part of the neighborhood trust community. Again, the trust community is distinguished by the heterogeneity of its membership.

The members of the trust community are active participants in the creation of its meaning. Just because a community is held together by viewing the evening newscast does not mean that the viewers believe the news reports. The opposite may be true—the viewers may be engaged in collective re-interpretation of the news if their prior experience suggests that the news as reported is unlikely to be true or complete. Then the news reporters and editors—aware of the audience—are implicated in this interpretation and re-interpretation. The common cause of this trust community is to interpret what is happening in the world. Even though they may diverge in their opinions, they are locked together in a community engaged actively in this work. This is the essential tension of a trust community; membership involves active interpretation of meaning.

This tension extends to the choices that the trust community makes about technology. What technology is used to hold together a trust community depends on what is available, popular, and suits the needs of the participants. It is a mistake to think that simply because a technology could enable a community of people to come together that it will. Just as members of a trust community participate in creating its meaning, they also participate in identifying which communications technology suits their purposes. The market is an important aspect of this process. When volunteers organized over the Internet choose what applications to use, they are likely to pick the simplest, most popular, most easily accessible ones available to them—ones that have already demonstrated success in the commercial market. Popular newscasts are politically important; they have large audience to interpret and perhaps reinterpret their information. Just because a television station transmitter is sending a signal into the ether does not make it part of a trust community. An evening newscast that no one watches is not significant as a trust community; there is no membership and no tension over its interpretation.

The power of the trust community is in its shared information and ideas—whether it is preventing flu epidemics, understanding the news, or improving traffic safety. These communities are not based on accumulation of land, of money, or of munitions. Information and ideas are their power base. The information that volunteers collect on prevention of flu epidemics are their primary resource in achieving their aims. The evening newscast is a brokered understanding among viewers and broadcasters, built across time, over what happened and what is reported. For neighbors trying to improve safety, the currency of exchange is knowledge of local streets, pedestrian behavior, and traffic rules. Therefore, when communications technologies reduce the effort and cost of exchanging information and ideas, new trust communities can emerge. People who had common cause but failed to act before may find that with new technologies they are able to communicate and engage together in changing the world around them. Using trust communities as an analytical lens can expand our understanding of information and ideas as a source of political power.

Using Technology to Lead

A Note to Activists, Businesses, and Governments

ONE OF THE pleasures of academic work is delving deeply into big questions, drawing on theory and history to examine a problem. The drawback is that sometimes readers find it challenging to link history and theory to current problems. Very often scholarly books leave to the reader's imagination the possible implications of research for practical application in politics. In this epilogue I try to speak briefly and plainly of the implications as I see them for three groups: activists seeking to change society, businesses with politically engaged customers, and governments seeking to maintain and expand their influence.

For the Activist: On Information as a Tool for Change

How relevant, really, are examples of protestors using telegrams to today's social activists? I suggest two sets of principles that can be drawn from these historical accounts. The first set applies to creating cohesion among your fellow activists; the second applies to the relationship of your campaign to the rest of society.

To create cohesion among activists, first invest, maintain, and improve your information cache continuously. Second, remember that your movement is defined by the quality and character of your information. Third, choose easy and flexible tools to collect and distribute information. Finally, keep in mind that the ways you handle information become the core customs and rituals of your group.

Think of information and ideas as your ammunition and capital; information is something that is worth investing in, maintaining, and seeking to improve. If yours is the sort of movement that is short on resources, then information and ideas may in fact be your principal tools. In the Philippines, the protestors supporting the ouster of President Estrada were not formally organized. The only common resource they had was the trust community created as an alternative to the new system Estrada sought to control. Their main capital was news. Similarly, with the *TsunamiHelp* blog and wiki, the volunteers' main resources were their computer equipment and Internet access. Using this equipment, their main activity was collecting and distributing information about who and what was happening when and where.

It follows that the quality and character of your information and ideas is very important to your work. It defines your movement. Whether your store of information is entertaining or deadly earnest, voluminous or incisive, innuendo or facts, its personality is the personality of the campaign. The *TsunamiHelp* volunteers had to decide what kind of information to collect. First, they were interested in news about what had happened. However, they also collected information about people's individual circumstances—missing, found, or dead. Further, they distributed information on fundraising efforts. All these decisions affected who participated and viewed the blog and wiki. It also led to a need to move off of a free platform, Wikinews, which was dedicated to news only and not to other kinds of information, and on to their own server. The volunteers lost some support and gained other support in the process of defining what information they would handle.

Use the communications tools that work best for the people in your community, not necessarily what is the latest fashion. Be aware of the communications companies on whom you depend for your services. In the 2011 Arab Spring, telephone companies, television broadcaster corporations, and social media firms directly affected whether activists could convey their messages to the world. Keep in mind that activists' campaigns frequently use a variety of media to connect participants and reach audiences. The Chinese public telegram protests relied on newspaper distribution to reprint the telegrams. In the Philippines, short messaging service (SMS) on cell phones was used to organize protests, but television, radio, and newspapers spread key background information on President Estrada's activities. In Taiwan, cable television news was preceded by distribution of political opposition videotapes. Cable television growth was part of a larger media liberalization that included a boom in newspapers and magazines. The International Campaign to Ban Landmines (ICBL) used not only email and Internet but also phones, faxes, print, and face-to-face meetings.

The activity of collecting and distributing information and ideas actually creates your community. You get to know your fellow activists by exchanging correspondence, discussing your cause, identifying who can help change society, and organizing for action. This creates opportunities for reciprocity and chances to build trust, share ideas and values, and create social capital. Throughout this process you are developing a repertoire of routines and rituals. How you communicate in ordinary times creates the habits your members will return to in times of crisis. In the ICBL, the use of regular reporting to create steady pressure on governments was a technique used by other organizations like Human Rights Watch. This was a novel technique to achieve, and subsequently to enforce, an international treaty. After the landmine campaign, similar efforts have been made for other causes like cluster bombs.

How you handle information among your fellow activists is as important as how you handle your finances or your equipment and supplies. The correspondence and exchange of ideas among you establishes routine patterns of behavior that people will naturally repeat in times of pressure or emergency. Choose the easiest and most flexible tools, and you will reach more people. The kind of information and ideas you trade gives your movement quality and character. If yours is a movement built by digging up previously unavailable information, you will become known for shining light on problems. If yours is a movement built on solving problems by accumulating mountains of data, you will become known as an expert organization available for consultation.

You can also consider using the communication routines and rituals of other activities in your own movement—basically, importing techniques from previous collective actions. For example, in China, Jing Yuanshan's public telegrams to help Emperor Guangxu against the Empress Dowager Cixi were not his first public telegram campaign. Earlier, he had tried unsuccessfully to build support for a codebook to use in the fight against Japan. Subsequent to the Guangxu matter, public telegrams were used for other issues as well. In the Philippines, the habit of political texting jump-started with the ouster of President Estrada. However, it continued afterward when Estrada's successor, Gloria Macapagal-Arroyo, used it to collect views and information from her supporters. Subsequently, when she was engulfed in an election scandal of her own, the texting turned against her when people used SMS and ringtones to ridicule and satirize her administration.

The second set of principles applies to the relationship of your movement with the rest of society. First, what is the sentiment, as yet unarticulated, that your movement will spark? Second, does your information compete or complement information already available? Third, who is your trust community—activists, observers, critics, opponents? Fourth, how does your trust community interpret your information?

There is still an indefinable something—a spark—that is necessary for a movement to take hold. Is there a "spiral of silence"[1] to be broken—something that is important to people but on which they are reluctant to take a public stand without the assurance that others are with them? The text messages and the *Pinoy Times* reports on President Estrada's corruption in the Philippines broke through his efforts to muzzle the media. In Taiwan, the liberalization of the media, and especially the explosion of political discussion on cable television, moved debates once confined to the dinner table out into the public square. The self-immolation of a street vendor defenseless against a corrupt government in Tunisia was the catalyst for Tunisians en masse to express years of pent-up frustration.

Be conscious of the goals that governments have—national identity, national security, and economic development. Are you pushing them in the direction they already profess, or are you challenging them in some way—by opposing or competing with them? For example, Al-Jazeera's decision to report news in a way that was more interesting to Arab viewers distinguished it from other government-owned Arabic-language broadcasts. It has gained a reputation for reporting facts, for discussing issues, and for addressing the concerns of the people of the Middle East. Al-Jazeera competes with other news agencies—some commercial, some public broadcasters—in presenting information. In contrast, the International Campaign to Ban Landmines built up a store of information that filled a gap in the world's knowledge. Previously, data and news about landmine victims, aid, and policy issues were scattered and uncoordinated. By pulling the arguments and information together, the campaign created a complementary well of information that expanded the foundation of knowledge used by governments to make policy decisions. As long as the campaign provided the information to the public, the governments could not claim ignorance.

How others interpret the information and ideas you distribute affects the success of your movement. Others may interpret your information and ideas differently than you intended. For example, India's public broadcaster Doordarshan intended the *Ramayana* television series to improve its popularity; the show succeeded in this, but with political implications not intended by the broadcaster. The BBC has built a reputation as a credible news agency and improved the Great Britain's standing in the world not by singing the praises of the British system but by accurately reporting news even when it was not favorable to Great Britain. Through how you handle your information, people will assess your intent and character, and through this lens will give meaning to the ideas you advocate.

Think of your trust community as broader than just your fellow activists. Others are part of your network as well, or there would be no cause to support. There are the people you are trying to persuade—how are they connected to you, in terms of infrastructure for communicating and the ideas you are seeking to convey? Then, once you have built your own activist group, how does your group relate to others? Do you compete or complement them in terms of membership, purpose, and—very critically—in the information and ideas you have at your disposal? There are the observers of your movement—is anyone watching? There are critics and commentators of your movement—they can bring you new volunteers or strengthen those you are seeking to change. In the *TsunamiHelp* example, journalists who reported on the volunteers' work helped them attract more recruits. It also attracted the attention of those raising funds, distributing relief materials, and reporting on the disaster. All these people were part of the trust community.

For Businesses: Your Customers' Trust Communities

Throughout this book, business and politics are intertwined. The main observation I draw is that if there is a conflict between government and activists, businesses can choose to take sides or to remain neutral, but all options require engagement. There is no shelter in inaction. When customers are part of trust communities, businesses are at least political participants and at most political leaders.

Businesses that are politically neutral can find their products and services adopted by activists. For example, in the Philippines, the government opened the mobile phone market to competition against the incumbent former telecom monopoly. Short message service (SMS) /texting service boomed and became a popular way to share news. As events coalesced around the ouster of President Estrada, customers used texts to organize. A popular service provided by private business, not the government, enabled the coup. Similarly, in the early 2000s, the commercially provided service *blogspot.com* was one of the most popular blogging platforms in the world. The initial *TsunamiHelp* volunteers took advantage of *blogspot.com* to launch their flagship blog. During the 2011 Arab Spring, Twitter and Facebook customers used these social media services to stay informed and act politically. In all these cases, the commercial service did not cause the social movement, but their success as popular services put them in a position to be used by customers for political purposes. If your business finds itself in this situation, ask: To what trust communities do your customers belong? With which trust communities do you want your business to engage? If these are not the same, how will you negotiate the differences?

Businesses can be leaders in activist movements. In the example of public telegrams in nineteenth-century China, Jing Yuanshan successfully ran the Shanghai telegraph company. He used his company as a resource to mobilize the public for political causes, some of which succeeded, while others failed. In Taiwan, the cable television operators started their businesses in the pursuit of profit. However, in addition to commercial success, they wanted a democratic government, and they used their clout toward that end. These businesses took advantage of the trust community they had built for commercial purposes and turned them toward political goals. If your business is at the core of a trust community of activists, then you face questions similar to activists with one added twist: Is there some way to integrate your goal of leading a trust community with growing your company? Are you converting customers to your trust community or converting your trust community into customers?

Businesses can work in concert with government goals. In Canada, the nation's transcontinental identity—promoted by the government—provided a catalyst for telecom companies to wire the nation. In the United States, commercial firms made the most of the early Internet technology funded by the government and showed how successful and popular it could really be. There are some government goals that are achieved most easily by business rather than government providing service directly. Two examples are the missed opportunity by the Soviet Union to develop an Internet and the troubles the Brazilian telegraph ran into when it was primarily a military development. Is there some way your business goals overlap with the government's?

Activists and government officials are also customers of businesses. The extent to which a business's customers are part of trust networks active in politics, the business is implicated. Just like other institutions, such as government, media, and social organizations, businesses are an integral part of the life and activity of trust communities.

For Governments: How Information and Ideas Shape the Nation

A nation is an idea held in the hearts of its people as much as it is a geographic unit ruled by a government. The examples in this book show the duality of government work. On the one hand, they are responsible for the practical, smooth functioning of communications infrastructure and services, and on the other hand, they influence the language, stories, and meanings that form the national identity. A government's soft power combines these two: (1) good infrastructure, which is used to deliver (2) influential ideas and information. From the historical examples in the book, several sets of lessons can be learned about how government can use information and ideas as tools to bring people together. These lessons are divided according to constituency—citizens, businesses, the discontented, and other nations.

For citizens, as technology improves, governments need to keep up with citizens' rising expectations of public services. Over the long run, governments can try to shape national identity, but ultimately the people interpret what it means to be a citizen. With regard to business, governments can promote the development of communications infrastructure and services; however, businesses are best at popularizing them. For those citizens who are discontented, governments can expect that activists will be early adopters of new technology. In response, suppressing information and ideas is possible but requires continuous effort. In considering the position of your government relative society, consider who is your trust community—government, businesses, citizens, opposition, other nations? And, how does your trust community interpret your information and ideas?

There are always basic services that governments are expected to provide. As technology improves, you need to keep up with citizens' rising expectations of public services. With new communications technologies available, the minimum expected standard will rise, and you should be prepared to meet it. The *TsunamiHelp* case is an example where a community came together because of the perceived failure of national and international institutions. Humanitarian organizations actually prepared as they always had for disasters, but they had not adapted to delivering services with new communications technology. Their slowness to adapt created an opportunity for the *TsunamiHelp* volunteers.

Some areas where this is particularly clear are disaster response and recovery, running elections, and regular release of government data like economic statistics, weather, and public health. These are all areas where people expect the government to deliver services in the timely, modern way that commercial services are also delivered. Not delivering opens up the government to possible challenge. If your citizens shop online, they will expect to file their taxes online as well. Governments that simply insist on their exclusive right to deliver such services—whether it is information like the weather, the economy, or public health, or services like telecom, television, or Internet—cannot expect such a defense to hold for long. E-government is not a special service; it is ordinary service delivered by modern technology.

The people are the ultimate interpreters of what it means to be a citizen. As for the information and ideas that go over the network, the better connected your nation is, the better informed people will be about the world. Geography is not identity—community is identity. If there is a national identity you are trying to foster, better information flows do create this opportunity for you. For example, in the early twentieth century, the Canadian government promoted the idea of the Canadian nation as bound by infrastructure running east-to-west across the continent—this included the telegraph, the telephone, and railways. The reasoning was that Canadians could easily communicate with each other and this would build community.

However, national identity is not simply a top-down creation that a government gives to the people. In Taiwan, the building of a cable television network unleashed in public an alternative, a local Taiwanese identity that had not appeared before in the government-controlled media. In India, the *Ramayana* program shown by public broadcaster Doordarshan similarly contributed to the rise of Hindu identity politics. People receive messages from the government in the context of all the other messages they receive and filter through their own worldview. Whether your message is successful will depend on whether people believe you, which depends on the totality of their experience with the government.

Government can influence technology development, but businesses often deliver the most popular services more quickly. Governments can start research, set high public aspirations, or take other actions which foster new technology. In Canada, the Diamond Jubilee was a celebration of a new national identity, a ceremony facilitated by broadcasting festive music and speeches over a newly connected communications infrastructure. By showcasing the importance of communications technology to the Canadian identity, the government gave commercial telephone operators an additional incentive to build their cross-country network, something they before had hesitated to complete. Similarly, you can identify policy goals to support the building and delivering of communications services to communities that are not commercially viable in order to make sure everyone is included.

The government also makes decisions about market regulation, which can affect how many businesses can enter the market to build communications infrastructure and provide services. In China, the government founded several state-owned telecommunications operators, and the rivalry among them has improved the coverage of service and lowered the price of phone service for ordinary people. In the Philippines, the decision to end the monopoly of the telecom operator and allow private businesses to offer mobile phone service triggered a boom in cell phone service and text messaging service, with direct implications for public protest and politics just a few years later. In general, the more participants in the market, the more competition there will be, and the more likely a wide range of communications services will be available to people at prices they can afford. This is true for infrastructure but also for media markets—where making and distributing popular ideas is the main commercial goal.

There is some danger in government intervening too much and preventing commercial innovation. In early-twentieth-century China, the Qing government-run telegraph network languished while the commercial telegraph run by Jing Yuanshan flourished. The Brazilian telegraph into the Amazon built by Rondon's military commission was not sustainable solely on government correspondence. Commercial messaging failed to take off. The Soviet government had advanced science and technology, but because controlling information flow was a higher priority, they did not explore the commercial opportunities offered by networking technology, and consequently those innovations took place in the United States and other countries. The exception has been in late-twentieth-century China, where the government managed competition among several state-owned telecom operators and successfully constructed a network across the nations. Rivalry among operators to meet growing demand after a decade of economic reforms kept development in line with need.

If there are activist groups, they will be early adopters of any new technology. This will create new opportunities for you to engage those constituencies. Jing Yuanshan's public telegrams in China in 1900 came about ten years after foreign-operated telegraphs began surreptitious operations around 1888.[2] In Taiwan, cable television came to political influence in the 1980s, about twenty years after the first cable television infrastructure appeared in the 1960s. In the Philippines, the protests against Estrada in 2001 used cell phones that became popular after regulatory liberalization in 1992; short messaging service used in mobilizing protestors was introduced in the Philippines in 1999. The *TsunamiHelp* blog and wiki took place in 2004, about a decade after the Internet went commercial in the United States. The Internet went commercial later in India, where the *TsunamiHelp* work originated; blogs and wikis followed. When activists adopt technology, the first activity will be to exchange views and information. If there is sharp, widespread dissatisfaction with the government, the advent of a new technology may make it easier for new trust communities to organize.

Suppressing information and ideas is possible but requires continuous effort. If there are constituencies that need to be suppressed, criminal organizations, for example, there is always technology available to do this. There is much documentation of how governments today filter Internet content, shut down social networking sites, and block television and radio signals to achieve public policy goals. Also, as long as there is a commercial and national security demand for tracking down individual Internet users, that technology will continue to improve as well.

When these techniques are used regularly to suppress information, however, people develop other alternative communications networks, often informally. In Taiwan, the government-controlled television news maintained sway for three decades. Things changed because of a combination of the cable television infrastructure and the launch of satellite television programming. In the Middle East, the sterility of earlier Arab news services presented an opportunity for Al-Jazeera. If you suppress information, be aware that you may create an information distortion. For example, it was easier for scientists to learn about foreign technical developments than to learn about other Soviet developments because of government-imposed communications barriers between Soviet research centers.

In short, while it is possible to restrict ideas and information either to limit the world's entry into your country or to influence your country's image abroad, to be effective, it has to be a constant campaign. Given that alternative sources can easily spring up, the campaign must not only be about projecting ideas and information but about persuading people to them as well.

Think expansively about who is in your government's trust community. It includes more than just the ministries and departments of your government. It includes the people whom those ministries and departments serve. It includes the media who observe and report on your performance. It includes critics and even enemies who spend their time and resources opposing you. It includes other nations—and their departments, ministries, citizens, and critics—as well as the degree to which they respond to your policies and actions.

The only people who are not part of your trust community are those who ignore you in the belief that you have nothing to do with them. This is the very dilemma some governments use infrastructure to try to bridge. Brazil tried to bring the people in the Amazon within their trust community by commissioning telegraph construction in their land. More successfully, the Canadian government encouraged the expansion of the telecommunications infrastructure to enhance the transcontinental, east-west identity of the nation.

Once you have identified who is in your trust community, how do they respond to the information and ideas you provide? Is their interpretation consistent or inconsistent with your intent? Then an interesting question arises. What is more important for you as a government—that the ideas you *articulate* are consistent with your values, no matter what other people may think of them, or that your ideas are *received and interpreted* with meaning that is consistent with your values? For example, the BBC and Al-Jazeera indirectly boost their governments' image by providing a good news service valued by their viewers and listeners. The boost comes from the excellence of the service, not from direct praise of the government. The ideas are received and interpreted in a way that is consistent with the goals of the British and Qatar governments, but the ideas articulated by the broadcaster may in fact be somewhat critical of the governments.

Governments have always used information and ideas to govern and shape their nations. The primary lesson for governments from studying trust communities is that the community is an interactive space. The information and ideas set out by governments is subject to interpretation by other members of the trust community; and these interpretations can be radically different from the government's intention. Therefore, when the government leads—by setting lofty goals for the building of infrastructure or encouraging the science necessary to create new services—it is important to think not only of the ideas delivered but also of how the ideas will be received. Governments that are concerned only about the delivery of messages will find themselves in shrinking trust communities—people will find them less and less relevant to their lives. Governments that manage to have their ideas and information well received by people will find their trust communities expanding and, consequently, their influence and power climbing.

NOTES

Chapter One: Trust Communities from the Telegraph to the Internet

1. Baark 1997, p. 55.

2. I am indebted to J. P. Singh for making the distinction between extending and transforming political activity.

3. Zhou 2006, pp. 39–103.

4. Khanfar 2010.

5. Anderson 1991, p. 39.

6. Ibid., pp. 132–38.

7. R. Gupta 2008.

8. See http://tsunamihelp.blogspot.com.

9. Williams 2000, pp. 87–88. DeChaine 2005, pp. 121–29.

10. Williams 2000, pp. 87–88.

11. Castells 1996, p. 37.

12. Ibid., p. 214.

13. Hou 2003, p. 186, n. 15.

14. Chin 1997, p. 83; Hou 2003, p. 189.

15. Chin 1997, p. 84.

16. Hou 2003, pp. 187–88.

17. Marx 1977, pp. 247–55. For a discussion of England, see pp. 877–95.

18. Tyler 2001, pp. 38–42.

19. Noelle-Neumann 1993.

20. Tyler 2001, pp. 38–42.

21. Ostrom and Walker 2003, pp. 28–35.

22. Putnam, Leonardi, and Nanetti 1993, pp. 167–76.

23. Putnam 2001, pp. 18–24.

24. Mueller 2010, pp. 41

25. Ostrom 2005, p. 26.

26. Ibid., p. 259.

27. Braman 2006, p. 25.

Chapter Two: Blogs, Wikis, and International Collective Action

1. Asia Development Bank 2009; Lay 2005.

2. Hitwise New York 2005; Hitwise UK 2005; Hitwise Australia 2005.

3. Griffin 2007; Di Maio 2005.

4. Tarrow 1998, *Power,* pp. 29–30.

5. Ibid., p. 133.

6. Coleman 2013.

7. Tarrow 1998, Fishnets, pp. 235–29.

8. Hanagan 1998, p. xix.

9. Tarrow 1998, *Power,* p. 113.

10. Passy 2003; McAdam 2003.

11. McAdam 2003, pp. 282–83.

12. Tarrow 1998, *Power,* pp. 164–67.

13. See Schiffer 2001, pp. 215–26, for a broader discussion of how, as technology develops, new ones emerge, but the old ones may persist for some time.

14. Both Sikkink's (2009) work and Brinkerhoff (2009, pp. 88–98) underscore that these kinds of networks are flexible and adapt; people voluntarily join and exit; information and learning hold the network together and give it meaning.

15. Coleman 2011.

16. Brinkerhoff 2009, pp. 88–98.

Chapter Three: Activists Challenge Institutions with Information Technology Networks

1. Spence 1990, 229–30.

2. Zhou 2006, 39–103.

3. Ibid., 67–69.

4. Ibid., 39–103.

5. Ibid., 70–73.

6. Ibid., 61–62.

7. Ibid., 62.

8. Ibid., 67–68.

9. As quoted by Vicente Rafael 2003, p. 402. See also Mangahas 2001.
10. Coronel 2003, p. 61.
11. Reid 2001, p. 778.
12. Coronel 2003, pp. 61–63.
13. Ibid.
14. Lee 1992.
15. Riedinger 1994, 143–44.
16. Kim 2003, pp. 498–501; Tiglao 1998.
17. Tiglao June 1999 and December 1999.
18. Lallana 2005, p. 10.
19. Ibid., p. 15.
20. Chin 1997, 82–3.
21. Hou 2003, p. 186, n. 15.
22. Chin 1997, p. 83 and Hou 2003, p. 189.
23. Chin 1997, pp. 80–81.
24. Hashimoto 1998, pp. 213–18.
25. Chen 2002, p. 42.
26. Rampal 1994, pp. 79–88.
27. Chiu and Chan-Olmstead 1999, p. 494.
28. Chin 1997, p. 84.
29. Hou 2003, pp. 187–88.
30. Chiu and Chan-Olmstead 1999, pp. 493–95.
31. Mekata 2000, p. 143; Hubert 2004, p. 78.
32. Mekata 2000, p. 145.
33. DeChaine 2005, p. 105.
34. A fax is a document sent over a telephone line. On each end of the telephone line a printer with a modem is attached. The modem converts the images on the paper into digital form, and the signal is sent over the telephone line. At the destination, the receiving fax machine's modem converts the digital signal back into images that are then printed onto paper.
35. Mekata 2000, pp. 143–45.
36. Williams 2000, pp. 87–88.
37. Ibid., and DeChaine 2005, pp. 121–9.
38. Williams 2000, pp. 87–88.
39. Beier 2003, p.795.
40. Hubert 2004, pp. 96–8.
41. Wareham 2008, pp. 49–54.

42. At www.the-monitor.org, see "history."

43. Wareham 2008, pp. 52.

44. Wareham 2008, pp. 49–54.

45. www.the-monitor.org/index.php/LM/Our-Research-Products/Landmine-Monitor.

46. Fathy 2011, p. 39.

47. Schrader and Redissi 2011, pp. 6–7.

48. Howard and Hussain 2011, p. 38.

49. Ibid., pp. 36–38.

50. Telegeography 2011.

51. Schrader and Redissi 2011, p. 11; Memm 2011.

52. Schrader and Redissi 2011, p. 11.

53. Howard and Hussain 2011, p. 39.

54. Etling, et al. 2009.

55. Schrader and Redissi 2011, p. 14.

56. Fathy 2011, pp. 39–34.

57. Noelle-Neumann, 1993.

58. Schiffer 2001.

Chapter Four: Governments Shape Nations with Communications Technology

1. "Canada Diamond Jubilee." http://archives.cbc.ca/programs/1364/.

2. Rens and Roth 2001, pp. 206–12.

3. Ibid. See http://archives.cbc.ca/programs/1364/ for the audio clip. See www.cn.ca/about/company_information/history/CNRadio5.htm for photos.

4. Rens and Roth 2001, pp. 156–57.

5. Diacon 2004, pp. 10–15.

6. Ibid., p. 9, 51, 115–29.

7. Ibid., pp. 56–71.

8. Ibid., pp. 138–56.

9. Winseck and Pike 2007, 44.

10. Winseck 1995, 2–3.

11. See http://go.worldbank.org/RTP5F5X7D0 (accessed October 10, 2010).

12. Wu 2009, pp. 14–20.

13. Yang 2009.

14. Wheeler 2006, pp. 143–44.

15. Ibid., pp. 10–11.
16. Ibid., pp. 143–44.
17. Ibid., pp. 142–43.
18. Ibid., pp. 27–28.
19. Brock 2002, 44–74.
20. Gerovitch 2001, p. 261.
21. Gerovitch 2008, pp. 344–46.
22. Dickson 1988, p. 1034.
23. Gerovitch 2008, pp. 344–46.
24. Cave 1980, p. 181, Gerovitch 2001, p. 56.
25. Goodman 1979, pp. 550–51.
26. Gerovitch 2001, p. 269.
27. Goodman 1979, pp. 564–65.
28. Gerovitch 2001, p. 273.
29. Goodman and McKenry 1991, p. 25.
30. Marker 1985, pp. 17–25.
31. Ibid., pp. 81–102.
32. Ruud 1990, pp. 17–29.
33. Ibid., pp. 29–35.
34. Ibid., p. 158.
35. Ibid., pp. 70–71.
36. Ibid., pp. 163–64.
37. Ibid., pp. 148–51
38. Ibid., p. 175.
39. Diebert 2008, pp. 9–19.
40. Ibid., pp. 294–95.
41. Galaz 2010, pp. 20–28.
42. U.S. Internal Revenue Service Oversight Board, p. 5.
43. Cortada, pp. 16–48. West, pp. 83–89.
44. U.S. Internal Revenue Service Oversight Board, p. 5.
45. West 2005, pp. 1–12.
46. Rajagopal 2001, p. 84.
47. Farmer 2005, p. 107.
48. Mankekar 1999, p. 165.
49. Rajagopal 2001, 84.
50. Gupta, N. 1998, 41.
51. Farmer 2005, 104.

52. Rajagopal 2001, p. 30.

53. Ibid., p. 205.

54. Mankekar 1999, p. 165.

55. Jebaraj 2011. I am indebted to Professor Kiran Prasad for sharing this article with me.

56. Clara Peller, the actress in the "Where's the beef?" TV ad, *New York Times,* August 12, 1987.

57. Noam 1991, 1–10.

58. See http://advertising.bbcworldwide.com/home/mediakit/reachaudience/bbcworldnews/bbcworldnewsafrica. Temin 2003, 654.

59. Barbour 1951, p. 59, Ayish 1991, p. 377.

60. Ayish 1991, p. 377.

61. For a fuller discussion, see Wring 2005.

62. Miles 2005, pp. 58, 16–18.

63. Horan 2010, pp. 9–16.

64. El-Nawawy and Iskander 2003, pp. 83–85; Khanfar 2010. Horan 2010, pp. 17–21.

65. Rabi 2009, p. 446.

66. Howard and Hussain 2011, p. 45.

67. See government archives of the most recent review of the BBC's Royal Charter at http://webarchive.nationalarchives.gov.uk/+/www.bbccharterreview.org.uk/publications/CR_PUBS/crpubs_home.html.

68. I am indebted to research by Pedro Davies for this insight.

Chapter Five: Technology+Trust=Political Influence

1. Ostrom and Walker 2003, p. 33.
2. Ostrom and Walker 2003, p. 381.
3. McAdam 2003, pp. 282–85.
4. Putnam, Feldstein, and Cohen 2003, p. 9.
5. Ostrom and Walker 2003, pp. 27–34.
6. Putnam, Feldstein, and Cohen 2003, p. 187.
7. Ibid., p. 291.
8. Tyler 2011, pp. 27–47.
9. Noelle-Neumann 1993, p. 22.
10. Ibid., pp. 61–62.
11. International Telecommunications Union 2002, pp. 12–15.
12. Anderson 1991, pp. 41–49.

13. Fahmy 2010, p. 83.
14. Ang 1985, 136.
15. Passy 2003, pp. 22–25.
16. Giugni and Passy 2001, pp. 123–53.
17. See Keohane and Nye 1998.
18. Sikkink 2009, pp. 229–30.
19. Bennett 2005.
20. See Putnam, Leonardi, and Nanetti 1993, pp. 173–75.
21. Castells 1998, pp. 380–84.
22. Benkler 2006, pp. 3–23.
23. Braman 2006, pp. 9–29.
24. Tilly 2010.
25. Sikkink 2009, p. 240.
26. Deutsch 1966, pp. 87–101.
27. Anderson 1991, pp. 15–16.
28. Steger 2008, p. 6.

Epilogue: Using Technology to Lead

1. Noelle-Neumann 1993.
2. Wu 2009, 13.

REFERENCES

Anderson, Benedict R. O'G. 1991. *Imagined communities: Reflections on the origin and spread of nationalism.* London; New York: Verso.

Ang, Ien. 1985. *Watching Dallas: Soap opera and the melodramatic imagination.* London; New York: Methuen.

Asia Development Bank. August 1, 2010. Tsunami response: Updated 23 September 2009. Available from www.adb.org/tsunami.

Ayish, Muhammed. 1991. Foreign voices as people's choices: BBC popularity in the Arab world. *Middle Eastern Studies* 27 (3): 374–89.

Baark, Erik. 1997. *Lightning wires: The telegraph and China's technological modernization, 1860–1890.* Westport, CT: Greenwood Press.

Barbour, Neville. 1951. Broadcasting to the Arab world: Arabic transmissions from the BBC and other non-Arab stations. *Middle East Journal* 5 (1): 57–69.

Beier, J. M. 2003. "Emailed applications are preferred": Ethical practices in mine action and the idea of global civil society. *Third World Quarterly* 24 (5) (10): 795–808.

Benkler, Yochai. 2006. *The wealth of networks: How social production transforms markets and freedom.* New Haven, CT: Yale University Press.

Bennett, W. Lance. 2005. Social movements beyond borders: Organization, communication, and political capacity in two eras of transnational activism. In *Transnational protest and global activism,* ed. Donatella della Porta and Sidney Tarrow, pp. 203–26. Boulder, CO: Rowman & Littlefield.

Braman, Sandra. 2006. *Change of state: Information, policy, and power.* Cambridge, MA: MIT Press.

Brinkerhoff, Jennifer M. 2009. *Digital Diasporas: Identity and transnational engagement.* Cambridge; New York: Cambridge University Press.

Brock, Gerald. 2002. Historical overview. In *Handbook of telecommunications economics,* ed. Martin Cave, Sumit Kumar Majumdar, and Ingo Vogelsang, pp. 44–74. Amsterdam; Boston: Elsevier.

Canadian Broadcast Corporation. Canada diamond jubilee 1927. Available from www.cbc.ca/archives/categories/society/celebrations/celebrations-general/diamond-jubilee-broadcast-links-canadians.html.

Castells, Manuel. 1996. *The rise of the network society.* Malden, MA: Blackwell Publishers.

Cave, Martin. 1980. *Computers and economic planning: The Soviet experience.* Cambridge; New York: Cambridge University Press.

Chen, Ping-Hung. 2002. The role of the state in shaping Taiwan's cable television industry. *Media Asia* 29 (1): 37–42.

Chin, S. 1997. Broadcasting and new media policies in Taiwan. In *Media in global context: A reader,* ed. Annabelle Sreberny, pp. 77–97. London: Arnold; New York: St. Martin's.

Chiu, Peilin, and Sylvia Chan-Olmstead. 1999. The impact of cable television on political campaigns in Taiwan. *International Communication Gazette* 61: 491–509.

Coleman, E. Gabriella. 2011. Anonymous: From the *lulz* to collective action. *The New Everyday* (April 6, 2011).

———. 2013. *Coding freedom: The ethics and aesthetic of hacking.* Princeton, NJ: Princeton University Press.

Coronel, Sheila. 2002. New media played a role in the people's uprising. *Nieman Reports* (Summer).

Cortada, James W. 2008. *The Digital hand,* Vol. III: *How computers changed the work of American public sector industries.* Oxford: Oxford University Press.

DeChaine, D. Robert. 2005. *Global humanitarianism: NGOs and the crafting of community.* Lanham, MD: Lexington Books.

Deibert, Ronald. 2008. *Access denied: The practice and policy of global Internet filtering.* Cambridge, MA: MIT Press.

Deutsch, Karl W. 1966. *Nationalism and social communication.* Cambridge, MA: MIT Press.

Diacon, Todd A. 2004. *Stringing together a nation: Candida Mariano da Silva Rondon and the construction of a modern Brazil, 1906–1930.* Durham, NC: Duke University Press.

Dickson, David. 1988. Glasnost: Soviet computer lag. *Science* 241 (August 26): 1034.

Di Maio, Paola. Tsunami help blog: Developers notes. February 2005. www.academia.edu/687821/Tsunami_help_blog_developers_notes_Experiences_and_lessons_learned.

El-Nawawy, Mohammed, and Adel Iskander. 2003. *Al-Jazeera: The story of the network that is rattling governments and redefining modern journalism.* Boulder, CO: Westview.

Etling, Bruce, John Kelly, Rob Faris, and John Palfrey. 2009. Mapping the Arabic blogosphere: Politics, culture, and dissent. *Berkman Center Research Publication* No. 2009–06, (6) (June 2009), http://cyber.law.harvard.edu/publications/2009/Mapping_the_Arabic_Blogosphere.

Fahmy, Z. 2010. Media-capitalism: Colloquial mass culture and nationalism in Egypt, 1908–1918. *International Journal of Middle East Studies* 42: 83–103.

Farmer, Victoria. 2005. Mass media: Images, mobilization, and communalism. In *Making India Hindu: Religion, community, and the politics of democracy in India,* ed. David E. Ludden, 98–115. Delhi: Oxford University Press.

Fathy, Basem. 2011. A "Cute" Facebook revolution? *Revaluation and Political Transformation in the Middle East: Agents of Change* 3 (August 20), www.mei.edu/content/cute-facebook-revolution.

Galaz, V. (2010). Pandemic 2.0. Can information technology help save the planet? *Environment: Science and Policy for Sustainable Development* 51(6): 20–28.

Gerovitch, Slava. 2001. "Mathematical machines" of the cold war: Soviet computing, American cybernetics, and ideological disputes in the early 1950s. *Social Studies of Science* 31 (2): 253–87.

———. 2008. InterNyet: Why the Soviet Union did not build a nationwide computer network. *History and Technology* 24 (4): 335–50.

Giugni, Marco, and Florence Passy. 2001. *Political altruism? Solidarity movements in international perspective.* Lanham, MD: Rowman & Littlefield.

Goodman, Seymour. 1979. Soviet computing and technology transfer: An overview. *World Politics* 31 (4): 539–70.

Goodman, Seymour, and W. K. McKenry. 1991. The soviet computer industry: A tale of two sectors. *Communications of the Association for Computing Machinery (ACM)* 34 (6): 25–29.

Griffin, Peter. 2007. A candle in my window. In *Communicating disaster: An Asia Pacific Resource Book,* ed. Nalaka Gunawardene and Frederick Noronha, pp. 13–20. Bangkok: United Nations Development Programme.

———. 2008, February 28. Interview in Mumbai, India. See also www.caferati.com and zigzackly.blogspot.com.

Gupta, Nilanjana. 1998. *Switching channel: Ideologies of television in India.* New Delhi: Oxford University Press.

Gupta, Rohit. 2008, March 1. Interview in Jaipur, India. See also algomantra .blogspot.com.

Hanagan, Michael P., Leslie Page Moch, and Wayne Ph Te Brake. 1998. *Challenging authority: The historical study of contentious politics.* Minneapolis: University of Minnesota Press.

Hashimoto, Hidekazu. 1998. The importance of the free circulation of information. *Media Asia* 25 (4): 213–18.

Hitwise Australia Newsletter. Australians eager to help. 2005 [cited August 1, 2010]. Available from www.hitwise.com/news/au200501.html

Hitwise Great Britain. Asia disaster prompts surge in traffic to charity websites, December 30, 2004. [cited September 13, 2010]. Available from hitwise.co.uk.

Hitwise New York. Tsunami blog among 10 most popular humanitarian sites: Traffic to humanitarian sites up by 180% January 5. [cited January 5, 2005].

Horan, Deborah. 2010. Shifting sands: The impact of satellite TV on media in the Arab world. *Center for International Media Assistance* (March 29), cima.ned.org/events/past-events/2010-events/shifting-sands.

Hou, Cheng-nan. 2003. *The transition of alternative media in Taiwan, 1970s–2002: A historical, political, and sociological examination.* Buffalo: State University of New York.

Howard, Philip, and Muzammil M. Hussain. 2011. The role of digital media: The upheavals in Egypt and Tunisia. *Journal of Democracy* 22 (3) (July): 35–48.

Hubert, Don. 2004. "New" humanitarian advocacy: Civil society and the landmines ban. In *Fighting for human rights,* ed. Paul Gready. London; New York: Routledge.

International Telecommunications Union. 2002. Pinoy Internet: Philippines case study.

Jebaraj, Priscilla. 2011. The richness of the *Ramayana,* the poverty of a university: Interview of historian Romila Thapar. *The Hindu,* October 28, 2011.

Keohane, Robert, and Joseph Nye. 1998. Power and interdependence in the information age. *Foreign Affairs* 77(5): 81–94.

Khanfar, Wadah. 2010. Remarks at conference, Information in the Arab World, Georgetown University. March 23, 2010.

Khanwalkar, Seema. 2008, March 5. Interview in Ahmedabad, India.

Kim, Dong-Yeob. 2003. Economic liberalization and the Philippine telecom industry. *Journal of Contemporary Asia:* 493–509.

Lallana, Emmanuel C. SMS and democratic governance in the Philippines. *Report of the Commissioner on Information and Communications Technology, Office of the President, Republic of the Philippines* Version 2.0, www.apdip.net/projects/e-government/capblg/casestudies.

Lay, Thorne. 2005. The great Sumatra Andaman earthquake of 26 December 2004. *Science* 308: 1127–33.

Lee, Kuan Y. Speech at the Philippines business conference, November 19, 1992; Manila, Philippines. [cited October 21, 2010]. Available from stars.nhb.gov.sg/stars/tmp/lky19921118.pdf.

Mangahas, Malou. 2001. Text messaging comes of age in the Philippines. *Reuters Technology News,* January 28.

Mankekar, Purnima. 1999. *Screening culture, viewing politics: An ethnography of television, womanhood, and nation in postcolonial India.* Durham, NC: Duke University Press.

Marker, Gary. 1985. *Publishing, printing, and the origins of intellectual life in Russia, 1700–1800.* Princeton, NJ: Princeton University Press.

Marx, Karl, and Ben Fowkes. 1977. *Capital: A critique of political economy.* New York: Vintage Books.

McAdam, Doug. 2003. Beyond structural analysis: Toward a more dynamic understanding of social movements. In *Social movements and networks: Relational approaches to collective action,* ed. Mario Diani and Doug McAdam. Oxford; New York: Oxford University Press.

Mehta, Dina. 2008, February 28. Interview in Mumbai, India. See also dinamehta.com and mosci.com.

Mekata, Moto. 2000. Building partnerships toward a common goal: Experiences of the international campaign to ban landmines. In *The third force: The rise of transnational civil society,* ed. Ann Florini. Tokyo; Washington, DC: Carnegie Endowment for International Peace.

Memm, Joseph. 2011, September 25. The scary face of hactivism. *Financial Times,* Life and Arts sec., 1–2.

Miles, Hugh. 2005. *Al-Jazeera: The inside story of the Arab news channel that is challenging the west.* New York: Grove Press.

Monteiro, Anjali, and K. T. Jayasankar. 2008, February 28. Interview in Mumbai, India.

Mueller, Milton. 2010. *Networks and states: The global politics of internet governance.* Cambridge, MA: MIT Press.

Noam, Eli. 1991. *Television in Europe*. New York: Oxford University Press.

Noelle-Neumann, Elisabeth. 1993. *The spiral of silence: Public opinion, our social skin*. Chicago: University of Chicago Press.

Ostrom, Elinor. 2005. *Understanding institutional diversity*. Princeton, NJ: Princeton University Press.

Ostrom, Elinor, and James Walker. 2003. *Trust and reciprocity: Interdisciplinary lessons from experimental research*. New York: Russell Sage Foundation.

Passy, Florence. 2003. Social networks matter: But how? In *Social movements and networks: Relational approaches to collective action*, ed. Mario Diani and Doug McAdam. Oxford; New York: Oxford University Press.

Putnam, Robert D. 2001. *Bowling alone: The collapse and revival of American community*. New York: Simon & Schuster.

Putnam, Robert D., Lewis M. Feldstein, and Don Cohen. 2003. *Better together: Restoring the American community*. New York: Simon & Schuster.

Rabi, Uzi. 2009. Qatar's relations with Israel: Challenging Arab and gulf norms. *Middle East Journal* 63 (3): 443–59.

Rafael, Vicente. 2003. The cell phone and the crowd: Messianic politics in the contemporary Philippines. *Public Culture* 15 (3): 399–425.

Rajagopal, Arvind. 2001. *Politics after television: Religious nationalism and the reshaping of the Indian public*. Cambridge; New York: Cambridge University Press.

Rampal, Kuldip R. 1994. Post-martial law media book in Taiwan. *International Communication Gazette* 53: 73–91.

Reid, Ben. 2001. The Philippine democratic uprising and the contradictions of neoliberalism: EDSA II. *Third World Quarterly* 22 (5) (10): 777–93.

Rens, Jean-Guy, and Kaäthe Roth. 2001. *The invisible empire: A history of the telecommunications industry in Canada*. Montreal: McGill–Queen's University Press.

Riedinger, Jeffrey. 1994. The Philippines in 1993: Halting steps toward liberalization. *Asian Survey* 34 (2) (February): 139–46.

Ruud, Charles A. 1990. *Russian entrepreneur: Publisher Ivan Sytin of Moscow, 1851–1934*. Montreal; Buffalo, NY: McGill–Queen's University Press.

Schiffer, Michael B. 2001. *Anthropological perspectives on technology*. Albuquerque: University of New Mexico Press.

Schrader, Peter, and Hamadi Redissi. 2011. Ben Ali's fall: The upheavals in Egypt and Tunisia. *Journal of Democracy* 22 (3): 5–19.

Sikkink, Kathryn. 2009. The power of networks in international politics. In *Networked politics: Agency, power, and governance,* ed. Miles Kahler. Ithaca, NY: Cornell University Press.

Singh, J. P. 2011. *Globalized arts: The entertainment economy and cultural identity.* New York: Columbia University.

Spence, Jonathan D. 1990. *The search for modern China.* New York: Norton.

Steger, Manfred B. 2008. *The rise of the global imaginary: Political ideologies from the French revolution to the global war on terror.* Oxford; New York: Oxford University Press.

Tarrow, Sidney G. 1998. Fishnets, internets, and catnets: Globalization and transnational collective action. In *Challenging authority: The historical study of contentious politics,* ed. Michael P. Hanagan, Leslie Page Moch, and Wayne Ph Te Brake. Minneapolis: University of Minnesota Press.

———. 1998. *Power in movement: Social movements and contentious politics.* Cambridge; New York: Cambridge University Press.

Telegeography.com. Tunisia. [cited December 2011]. Available from telegeography.com.

Temin, Jonathan. 2003. Considering the role of the BBC in African conflict. *Review of African Political Economy* 30 (98): 654–60.

Tiglao, Roberto. 1998. Telecoms shake-out. *Far Eastern Economic Review* (December 17): 14.

———. 1999. Mixed messages. *Far Eastern Economic Review* (June 24): 64.

———. 1999. Telecoms turf war. *Far Eastern Economic Review* (December 9): 52.

Tilly, Charles. 2010. Cities, states, and trust networks. *Theory and Society* 39:265–80.

Turnbull, Susan. 2008, January 3. Interview in Washington, DC.

Tyler, Tom R. 2011. *Why people cooperate: The role of social motivations.* Princeton, NJ: Princeton University Press.

United States Internal Revenue Service Oversight Board. Electronic filing 2012: Annual report to Congress. December 2012. www.treasury.gov/irsob/reports/2013/IRSOB~E-File%20Report%202012.pdf (accessed January 10, 2014).

Wareham, Mary. 2008. Evidence-based advocacy: Civil society monitoring of the Mine Ban Treaty. In *Banning landmines: Disarmament, citizen diplomacy, and human security,* ed. Jody Williams, Stephen D. Goose, and Mary Wareham. Lanham, MD: Rowman & Littlefield.

West, Darrell M. 2005. *Digital government: Technology and public sector performance*. Princeton, NJ: Princeton University Press.

Wheeler, Tom. 2006. *Mr. Lincoln's T-mails: The untold story of how Abraham Lincoln used the telegraph to win the Civil War*. New York: Collins.

Williams, Jody. 2000. David with Goliath. *Harvard International Review* 22 (3) (Fall): 87–88.

Winseck, Dwayne. 1995. A social history of Canadian telecommunications. *Canadian Journal of Communications* (20) 2. www.cjc-online.ca/index.php/journal/article/view/863/769.

Winseck, Dwayne Roy, and Robert M. Pike. 2007. *Communication and empire: Media, markets, and globalization, 1860–1930*. Durham, NC: Duke University Press.

World Bank. Information and communications technologies. http://go.worldbank.org/RTP5F5X7D0 [cited October 10, 2010].

Wring, Dominic. 2005. Politics and the media: The Hutton inquiry, the public relations state, and the crisis at the BBC. *Parliamentary Affairs* (April): 380–93.

Wu, Irene S. 2008. Information, Identity, and Institutions: How Technology Transforms Political Power in the World. Yahoo! Fellowship paper. Institute for the Study of Diplomacy, School of Foreign Service, Georgetown University. http://isd.georgetown.edu/sites/isd/files/Yahoo_Wu_Information_Identity_and_Institutions.pdf [cited April 2014].

———. 2009. *From iron fist to invisible hand: The uneven path of telecommunications reform in China*. Palo Alto, CA: Stanford University Press.

Yang, Guobin. 2009. *The power of the Internet in China: Citizen activism online*. New York: Columbia University Press.

Zhou, Yongming. 2006. *Historicizing online politics: Telegraphy, the Internet, and political participation in China*. Stanford, CA: Stanford University Press.

INDEX

activists/activism, 16, 123–28; and business, 128, 129, 130; cohesion among, 123; and communications technology, 5, 20, 71; in Egypt, 67; and information, 134; and new technology, 130, 133–34; online, 112; relationship of, with society, 126–28; and technology, 5–6
Advanced Research Projects Agency (ARPA), 83, 84
advertising, 68, 69
Al-Jazeera, 6–7, 21, 70, 71, 98, 99, 113, 134, 135; and business and commerce, 97, 127; competition by, 127; and diversity, 111; and information, 127; *The Opposite Direction* talk show, 96; Satellite TV, 95–96
Amazon, 8, 79, 98, 133, 135
Anantula, Suhit, 29, 34, 36, 44
Anderson, Benedict, 7, 75, 76, 109, 115
Ang, Ien, 110
Anonymous (hactivist group), 50
Arabic news services, 21, 95, 134
Arab Spring, 124–25, 129
ARPANET, 83, 84
audience, 97; of BBC, 93–95; of Doordarshan's *Ramayana,* 91–93; and interaction, 104; interpretation by, 74, 89, 93, 110; and media capitalism, 109; participation of, 118; for public broadcasting, 99; response of, 118
Bahrain, 96
Beijing, 58
Ben Ali, Zine El Abidine, 67, 68, 71
Benkler, Yochai, 113
Bertolucci, Katherine, 29, 31–32, 34, 36, 39, 43, 44
Birth of Ram (Ram Janmabhumi), 92
Blair, Tony, 95
Blogger, 68
blogs/blogging, 9, 27, 29, 68, 70–71
blogspot.com, 29, 129
Bohrer, Nancy, 28, 34–36, 39, 41, 43–44, 45
Bolsheviks, 87
Bouazizi, Mohamed, 67–68, 127
Braman, Sandra, 113–14
Brazil: commerce in, 21, 77, 79–80, 98; government of, 8, 77–80; military in, 8, 78–79, 130; and national unification, 78; telegraph system in, 8, 21, 77–80, 98, 130, 133, 135
Brazilian Indian Protection Service, 78
Brinkerhoff, Jennifer, 50
British Broadcasting Corporation (BBC), 70, 96; Arabic service of, 94; and British identity, 94, 98, 113, 115, 127–28; and business and commerce, 97; and Hutton Inquiry, 95; reputation of, 93–95, 98, 111, 113, 127–28, 135; War Game documentary, 94–95
British Foreign Office, 94
broadcasting, public, 91–97, 98, 111

Burma/Myanmar, 87
business and commerce: and Al-Jazeera, 97, 127; and American computer development, 85; and American Internet development, 83–84, 98, 129; and American telegraph network, 83, 98; and BBC, 97; and Brazilian telegraph system, 21, 77, 79–80, 98; in Canada, 21, 77, 80, 129, 132; in China, 56, 57, 81; and Chinese telegraph network, 57, 72, 108; and communications infrastructure, 97; and communications technology, 7–8; customers of, 128, 129, 130; and Doordarshan's *Ramayana,* 91, 97; and Egyptian and Tunisian social media, 72; and government, 87, 88, 96–97, 129–30, 131, 132–33; and Internet, 103, 120; and national security, 83; and personal identity, 109; and Philippine cell phones, 72, 108–9; and Philippine newspapers, 108–9; in Philippines, 129–30; and politics, 7–8, 93, 104, 109, 117, 128–29; and social media, 68–69; and Soviet control of information, 133; and Soviet Internet networks, 85; and Taiwanese cable television, 72, 109; and technology development, 132; and telephone, 80; and trust in technology, 104

cable television. *See under* Taiwan
Cairo, Egypt, 69
Canada, 78; business and commerce in, 21, 77, 80, 129, 132; Diamond Jubilee of, 74, 77, 98; and infrastructure, 21, 77, 98, 113, 115, 132, 135; and national identity, 21, 77, 98, 113, 115, 129, 131–32, 135; telecommunication companies in, 129; telephone network of, 21, 77, 80, 113, 115, 129, 132
Canadian Broadcast Corporation, 74
capital, 12, 114
capitalism, 108–9
capitalists and workers, 11, 12
Carvin, Andy, 29
Castells, Manuel, 113
Catherine the Great, 86
Catholic Church, Roman, 59
cell/mobile phones: and business and politics, 72, 108, 128–29, 132–33; and China, 81–82; in Egypt and Tunisia, 69; expansion of resources through, 72; and International Campaign to Ban Landmines, 64; as new technology, 133; and other media, 125; and Philippine protests, 9, 20, 55, 59–61, 72–73, 108, 125, 126, 128–29, 133
Center for Diplomatic Missions, 29–30
China, 71–72; business and commerce in, 56, 57, 72, 81, 108; cell phones in, 81–82; civil society in, 55; commerce in, 57, 108; communications infrastructure in, 82–83; economic development in, 81, 82; economic reform in, 98; newspapers in, 5–6, 56–57, 125; public telegrams in, 5–6, 20, 55, 56–58, 64, 123, 125, 126, 129; and Taiwan, 61; telecommunications infrastructure in, 99; telecommunications market in, 132; telecommunications network in, 133; telegraph network in, 108, 129, 133; website blocking by, 87, 98
Chinese Community Party, 61
Cilibrasi, Rudi, 32, 37, 38, 39–40, 41, 43, 45
citizens, 7, 118, 130–32, 135
Cixi, Empress Dowager, 5, 6, 56, 57, 58, 126
Coleman, Gabriella, 50
collective action, 71, 126; and communications technology, 5;

and Internet, 105; and social capital, 105, 108; and social media, 117; and trust, 14, 15, 23, 103; and *TsunamiHelp,* 23, 26, 43, 46–47, 48
commonalities, 119; and communities, 15; and *TsunamiHelp* volunteers, 118
common cause: and communications infrastructure, 7, 103–4, 119, 121; and community, 10, 15, 108; and diversity, 111; and engagement, 109–10; and ideas as power source, 113; and interpretation, 120; and national purpose, 116; and *TsunamiHelp,* 26, 30, 39
communication: computerized, 106; and cooperation, 105; interactive nature of, 74; and International Campaign to Ban Landmines, 64; and nationhood, 75; online vs. face-to-face, 5. *See also* face-to-face communications
communications companies, 124–25
communications infrastructure, 99, 117; and cable television in Taiwan, 109; and Canadian national identity, 21, 77, 98, 113, 115, 129, 132; in China, 82–83; and common cause, 7, 103–4, 119, 121; and Deutsch, 114–15; and national identity, 77–80; scope of, 97; and trust community, 115
communications networks, 20; and information policy, 75; and infrastructure policy, 75; interaction of, 88–89
communications technology, 108, 117; and activism, 5–6, 11, 20, 71, 133–34; and business and commerce, 7–8, 132; and communities, 16; expansion of resources through, 72; and government, 5, 74–99, 132; governments' control of, 20, 21, 134; human beings connected by, 11; and information and ideas, 5; innovation in, 16, 20–21; and institutions, 16; and interaction, 104; and International Campaign to Ban Landmines, 10, 64, 65; and national identity, 77–80, 91–97; and national security, 80–87; networks possible through, 16; and new trust communities, 20–21; and politics, 5, 16, 104, 108; and power, 55; and public services, 87–91; and reciprocity, 105, 116–17; and sense of self, 4; and social capital, 22, 108, 117; in Soviet Union, 84–85; and state, 20; suitability of, 120; and trust, 22, 103–4, 108, 116–17
Communist Party, Soviet, 84, 87
community, 15; cause and identity of, 10; and communications technology, 5, 16; connectedness of, 12–14; and Deutsch, 114–15; and global imaginary, 115; as identity, 131; and identity and trust, 16; imagined, 115; as interactive space, 135–36; and Internet, 103; as larger than nations, 76; network as, 3–4; online, 39, 106, 118; and print capitalism, 109
computers, 84–85, 90, 106, 113
Confederacy, 82, 83, 98
Craigslist, 106
Cruz, Anna Lissa, 32, 34, 37, 38, 41, 43, 44, 45

Dallas (television series), 110
democracy, 12, 61–63, 89, 107, 114, 129
Democratic Progressive Party (Taiwan), 12, 62
Deutsch, Karl, 75, 76, 114–15
Diacon, Todd, 78, 79
Di Maio, Paola, 29, 30–31, 40
diversity, 9–10, 26, 104, 111–13, 118, 119
Doordarshan, 21, 91–93, 97, 98, 127, 132

economic class, 10, 104, 105
economic development, 6, 75, 80–81, 82, 97, 98, 127
economic power, 20, 105
economy, 11, 79, 84
egalitarianism, 38–40, 49–50
e-government, 89, 91, 131
Egypt: government of, 67, 68; protests in, 20, 55, 67–71, 72, 106–7; revolution in, 96; social media in, 68, 69–70, 72, 106–7
Egyptian Youth Movement of 2004–6, 71
email, 14, 109; and American Internet development, 83; and Estrada, 59; and International Campaign to Ban Landmines, 10, 20, 64, 73, 125; and *TsunamiHelp*, 26, 48
Embuldeniya, Angelo, 31, 36, 42
Estrada, Joseph, 20, 59, 133; and business, 108–9; corruption of, 8–9, 55, 58, 72, 109, 126–27; intimidation of media by, 59; and mobile phones, 124, 125, 126, 128–29; ouster of, 9, 58, 59, 128
ethnicity, 10, 104, 105, 118
Etling, Bruce, 70

Facebook, 55, 67, 68, 69, 71, 129
face-to-face communications, 14, 106, 117; and International Campaign to Ban Landmines, 10, 64, 73, 125; and Internet, 46; online vs., 5; and *TsunamiHelp*, 48
fax, 10, 20, 64, 73, 125, 139n34
France, 33

Garcilliano, Virgilio, 61
Gebauer, Thomas, 64
Gerovitch, Slava, 84
Ghonim, Wael, 67
Giugni, Marco, 110–11
global imaginary, 76, 115
globalization, 76, 116
Global Public Health Intelligence Network, 89
globalvoices.org, 29
Globe Telecom, 60–61, 108
Glushkov, Viktor, 84, 85
GoDaddy, 37
Google, 29, 68, 69
Goose, Steve, 66
government(s), 16, 130–36; and BBC, 93–95; and Brazilian telegraph system, 8, 77–80; and business and commerce, 87, 88, 96–97, 128, 129–30, 132; and citizens, 130–32; and commitment, 114; and communications control, 20, 21, 134; and communications infrastructure and services, 130; and communications interactions, 74, 88–89; and communications policy, 75, 97; and communications technology, 5, 72, 74–99, 132; constituencies of, 130, 131; and content policy, 118; control of social networking by, 134; and discontent, 130–31; and economic development, 6; goals of, 97, 127; and information and ideas, 11, 75, 76, 87–90, 135, 136; and information and ideas control, 87, 88–89, 93, 95; and information and ideas interpretation, 6, 89, 93, 131; and information and ideas suppression, 131; and infrastructure, 75, 76, 80, 117; and International Campaign to Ban Landmines, 10, 64–65, 67, 112; and market regulation, 132–33; and national identity, 6–7, 130; and national security, 6, 88; and other nations, 130, 131, 135; and public broadcasting, 96–97; release of data by, 131; reporting to create pressure on, 125; services of, 89, 130–32; soft power of, 130; trust community of, 135; and *TsunamiHelp*, 24, 26, 28, 30, 32, 40, 41–42, 47; and universal service programs, 80; website blocking by, 87. *See also* nation; *specific nations*

Graham, George P., 74
Grant, Ulysses S., 82, 83
Greenpeace, 38
Green Team, 73
Green Team videos, 62, 63
Green Television Station, 62, 63
Griffin, Peter, 27, 28–29, 30, 34, 38, 40, 44
Guangdong, 58
Guangxu, Emperor, 5, 6, 56, 57, 58, 126
Guardian, 29
Gulf States, 6, 95
Gupta, Rohit, 28–29, 31, 34, 38, 40–41, 43

Hamad bin Khalifa Al Thani, 6–7, 96
Hangzhou, 58
Hawaii, 88
health, public, 78, 88
Hindu identity politics, 132
Hindu nationalist movement, 92
H1N1 flu virus, 89
Hong Kong, 58
Howell, James A., 29–30
humanitarian groups, 26, 29, 42, 131
humanitarianism, 28, 32, 57
human rights organizations, 112
Human Rights Watch, 66, 125
Hurricane Katrina, 41, 44–46, 88

identity, 13, 108; as binding communities together, 12–13; collective, 43; common, 109, 111, 113, 119; and communities, 10, 16, 131; emotional, 13; and institutions, 16; and media capitalism, 109; and networks, 16; and politics, 91, 132; shared, 7, 103; social, 13; and *TsunamiHelp,* 26, 42, 43, 46–47, 107. *See also* national identity; self, sense of
Imagined Communities (Anderson), 7, 115
India, 23, 44, 88; and bubonic plague outbreak in Surat, 89; and Doordarshan's *Ramayana,* 21, 91–93, 97, 98, 127, 132; Internet in, 134; and *TsunamiHelp,* 9, 23
Indians (Brazilian), 8, 78–79
Indonesia, 88
industrial revolution, 11, 12
information and ideas: and activism, 123–26, 134; and Al-Jazeera, 6, 71, 96, 98, 99, 107, 111, 113, 127; as ammunition and capital, 10–11, 99, 104, 124; and BBC, 111, 127–28; change influenced by, 4; collection and distribution of, 125; competing channels of, 20; as consistent with values, 135; customs and rituals for handling, 124, 125; decentralization of, 113; as defining movements, 123, 126; distribution of, 123; and government, 11, 75, 76, 87–90, 135, 136; and government and interpretation, 6, 89, 93, 131; government control of, 87, 88–89, 93, 95; and Hurricane Katrina, 88; and International Campaign to Ban Landmines, 10, 64–65, 66–67, 72, 73, 107, 127; and Internet, 4; interpretation of, 6, 9, 22, 74, 89, 92, 93, 104, 110, 120, 126, 127, 131, 135, 136; and national identity, 131–32; in network society, 113; as political currency, 10–11; and politics, 5; and power, 12, 20, 22, 48, 55, 73, 104–5, 113–14, 120–21; as shaping nation, 130–36; Soviet control of, 113, 133, 134; suppression of, 131, 134; and World Bank, 81. *See also* *TsunamiHelp* blog and wiki
instant messaging, 48
institutions, 13, 15, 48; and adaptation to trust communities, 117; and communications technology, 16; effect of new technologies on, 4–5; evolution of trust community into,

institutions (*cont.*)
55, 117; hierarchical, 103; and identity, trust, and social capital, 16; and *TsunamiHelp*, 26, 50
interaction, 106, 111; and audience, 104; and communications, 74; of communications networks, 88–89; face-to-face vs. mediated, 117; and government memes, 93; in institutions, 15; and Internet, 91, 103; and meaning creation, 8–9; and networks, 3; online, 117; and public broadcasters, 93; and reciprocity, 13–14, 117; and technology, 104, 110; and trust, 13–14, 48, 99, 104, 117
International Campaign to Ban Landmines (ICBL), 21, 55, 63–67, 112; and communications technology, 10, 20, 64, 65, 73, 125; and face-to-face communications, 10, 64, 73, 125; and information and ideas, 10, 64–65, 66–67, 72, 73, 107, 127
International Federation of Red Cross and Red Crescent, 9, 24
Internet, 48, 50, 80, 93, 99, 107, 109; American development of, 83–84, 98, 129; and BBC, 95; change influenced by, 4; and China, 82; and collective action, 105; and commerce, 83–84, 85, 98, 103, 120, 129; and e-government development, 91; in Egypt, 69–70; and Estrada, 59; expansion of resources through, 72; and face-to-face communications, 46; and governments, 74, 134; in India, 134; and information, 4; and interaction, 91, 103; and International Campaign to Ban Landmines, 64, 67, 125; and IRS, 90; and politics, 4, 16, 103, 108; search engines on, 68; and Soviet Union, 130; and state, 16; and trust, 103; and *TsunamiHelp*, 20; in Tunisia, 69, 70; and website blocking, 87, 98
Iran, 87, 98
Iraq war, 70, 95
Isvestia, 87

Japan, 58, 88, 126
Japan-Russia War, 86
Jing Yuanshan, 5–6, 56–58, 64, 71–72, 108, 126, 129, 133

Kala Ghoba art festival, 44
Katrinahelp, 45
Kelly, David, 95
Kline, Rob, 34, 41, 45

Landmine Monitor Group, 66–67
landmine treaty, 10, 20, 63–67, 125. *See also* International Campaign to Ban Landmines
Land Monitor Group, 72
Lenin, V. I., 87
Lévi-Strauss, Claude, 77
Lincoln, Abraham, 82–83, 98

Macapagal-Arroyo, Gloria, 58, 61, 126
Mackenzie, William Lyon, 77
Malaysia, 58
Manghas, Mahou, 58
Manila Times, 59
Marcos, Ferdinand, 58
Marx, Karl, 11, 12
McAdam, Douglas, 46–47, 48, 105
media: in Egypt, 69; and Estrada, 59; and identity, 109; manipulation of, 113; older vs. newer, 72–73; online, 110; and trust community of government, 135; and *TsunamiHelp*, 28, 29, 42, 47
medico international, 64
Mehta, Dina, 31, 42; and blog redesign, 34; and Hurricane Katrina, 45; and Indian government ban on blogs, 44; initial efforts of, 28, 29; as leader, 38;

motivation of, 32; and *Tsunami-Help* ethos, 39, 40; and wiki creation, 37
microblogs, 68
Middle East, 76, 94, 95, 96, 97, 99, 107, 113, 134
military, 20, 97–98, 105, 114; and American computer development, 85; and American Internet development, 83–84; and Brazilian telegraph system, 8, 78–79, 130, 133; and Civil War in US, 82–83; and International Campaign to Ban Landmines, 112; and Internet in Soviet Union, 85, 130
Mine Ban Treaty, 66. *See also* landmine treaty
Mondale, Walter, 93
Morquendi, 41
Mubarak, Hosni, 67, 68, 69
Mueller, Milton, 15
Muller, Bobby, 64
Murthy, Megha, 28, 42; and blog redesign, 33–34; and decision making, 39; and Hurricane Katrina, 45; later collective action by, 44; motivation of, 31; and *TsunamiHelp* ethos, 40; and wiki creation, 36, 37

narratives, 106, 115, 116, 117
nation, 105, 115–16; and communication, 75; communications as tying together, 77–80; and communications networks, 75; communities as larger than, 76; as idea, 7, 109, 130; as imagined community, 75, 115; information and ideas as shaping, 130–36; limited and sovereign, 115. *See also* government(s); state
national identity, 75, 97, 98–99, 114; and Anderson, 115; and Brazil, 77–80; and Canada, 21, 77, 98, 113, 115, 129, 131–32, 135; and China, 81; and communications infrastructure, 21, 77–80, 98, 113, 115, 129, 132; and communications technology, 77–80, 91–97; and Doordarshan's *Ramayana,* 91–93; and government, 6–7, 127, 130; and India, 21; and information and ideas, 131–32; and newspapers, 7; and Taiwan, 132
national imaginary, 76
Nationalist Party (Taiwan), 12, 61, 63
national security, 6, 75, 81, 82–83, 88, 97–98, 127
nation state, 11, 48
networks, 13, 14–16, 43, 103, 109; of civic engagement, 14; and communications technology, 16; as communities, 3–4; and decentralization, 113; engagement as creating meaning of, 110; ethnic, 114; heterogeneous, 112; and identity, 16; information in, 113; and International Campaign to Ban Landmines, 64, 66; of people, 64, 66; power in purpose of, 114; technology-mediated, 48; and trust communities, 15; and *TsunamiHelp* volunteers, 49; voluntary nature of, 112
newcomfarm.com, 44
newspapers, 7, 119; and Anderson, 115; and capitalism, 109; and Chinese public telegram protests, 5–6, 56–57, 125; and Estrada, 59, 108–9; and national identity, 7; in Philippines, 73, 125; readers' letters to, 110; in Russia, 86–87; in Taiwan, 73, 125; and *Tsunami-Help,* 26, 29, 42, 48
Ninth Asian Games, 92
Noelle-Neumann, Elisabeth, 107
nongovernmental organizations (NGOs), 10, 55, 63, 64, 66, 73
nongovernment institutions, 30
North Korea, 87

Olympics, 116
online activism, 112

online communications, 5
online communities, 39, 106, 118
online interactions, 117
online media, 110
online support groups, 3
Ostrom, Elinor, 13, 105, 106

Pakistan, 44, 87, 98
Palestine, 94
Paraguay, 8
Passy, Florence, 48, 110–11
Peter the Great, 86
Philippine Daily Inquirer, 59
Philippine Islands, 124, 126–27; business and politics in, 72, 108, 128–29, 132–33; cell phones in, 9, 58–61, 108, 125, 128–29, 133; corruption in, 8–9, 20, 55; media in, 125; protests against Estrada in, 20; Short Messaging Service in, 9, 20, 58, 60–61, 72–73, 108, 125, 126, 128–29
Philippines Long Distance Telephone (PLDT), 59, 60
Pinoy Times, 59, 126–27
Pitchandi, Bala, 31, 32–33, 34, 41, 42; changed life of, 44; and Hurricane Katrina, 45; initial response of, 29, 30; as leader, 37–38; and *quakehelp.info,* 44; and *TsunamiHelp* ethos, 40; and wiki creation, 36–37
politics, 117; and business and commerce, 7–8, 93, 104, 109, 117, 128–29; citizen engagement in, 118; and commerce, 109; and commercial development of technology, 104; and communications technology, 5, 7–8, 16, 20; and competing channels of information, 20; and distribution of power, 5; and Doordarshan's *Ramayana,* 91–93; and economics, 11; and identity, 91, 132; and information and ideas, 5; information as currency for, 10–11; and Internet, 4, 16, 103, 108; and interpretation of information and ideas, 22; and nation as imagined community, 115; in Philippine Islands, 72, 108, 126, 128–29, 132–33; and power, 11, 16, 26; in Taiwan, 11–12, 61–63, 73; and technology, 104, 108; and *TsunamiHelp* volunteers, 26
ProMED, 89
propaganda, 88–89, 94, 104
public broadcasting, 91–97, 98, 111
public health, 78, 88
public security, 87–90. *See also* national security
Pushkin, Alexander, 86
Putnam, Robert D., 14, 106, 112

Qatar, 135; censorship in, 95–96; funding by, 97; political influence of, 21, 99, 111; reputation of, 6–7, 98, 113; stature of, 21. *See also* Al-Jazeera
Qing dynasty, 6, 20, 55, 56, 57, 133
quakehelp.info, 44

radio, 26, 48, 73, 74, 77, 98, 109, 125, 134
railways, 80, 132
Ramayana (television series), 21, 91–93, 97, 98, 127, 132
Ramos, Fidel, 60
Rasputin, Grigori, 86–87
reciprocity, 8, 72; and activism, 125; and communications technology, 5; and cooperation, 105; and interaction, 13–14, 117; and networks, 3; and technology, 116–17; and trust, 14, 103, 104, 105; and *TsunamiHelp* volunteers, 26, 47
Red Cross, 4, 45
RedHerring.com, 30
religion, 10, 11, 104, 105, 118
Reuters AlertNet, 9, 23
Rondon, Candido Mariano da Silva, 8, 78, 133

Rotary International, 42
Russia, 76, 86–87
Russian Word, 86–87

Said, Khaled, 67, 70
satellite television, 21, 76, 134
Saudi Arabia, 96
security, public, 87–90. *See also* national security
self, sense of, 4, 13, 107, 114. *See also* identity
Shanghai, 57, 129
Shanghai Telegraph Administration, 5–6, 56
Sherman, Robert, 82, 83
Short Messaging Service (SMS): and Estrada ouster, 58; and Philippine protests, 20, 125, 126, 128–29; in Philippines, 9, 20, 60–61, 72–73, 108, 125; and *TsunamiHelp* volunteers, 36, 41, 42, 48
Sikkink, Kathryn, 112, 114
silence, spiral of, 107, 126
Singapore, 58
social capital, 14, 105, 108; and activism, 125; as binding communities together, 12–13; bridging vs. bonding, 14, 112; and communications technology, 5, 22; and diversity, 112; and institutions, 16; and norms, 14; and online communities, 106; and technology, 108, 117; and trust, 22, 103; and *TsunamiHelp* volunteers, 26, 46, 49
social justice, 112
social media, 9, 14–15, 109; and Al-Jazeera, 71; and Arab Spring, 125, 129; and business and commerce, 68–69; in Egypt and Tunisia, 68, 69–70, 72, 106–7; and narratives, 106; transactional interactions of, 117
social networks, 48, 134
socioeconomic class, 11, 111, 118
soft power, 105, 111, 130
South and Southeast Asia, 9–10, 23, 88
South-East Asia Earthquake and Tsunami blog, 28
South Korea, 87, 98
Soviet Communist Party, 84, 87
Soviet Union: communications technology policy in, 84–85; and computers, 84–85, 113; information control in, 113, 133, 134; and Internet, 130; military in, 85, 130; Politburo, 85; universal information bank in, 84–85
Sri Lanka, 88
state, 7, 16, 20. *See also* government(s); nation
Steger, Manfred, 75–76, 115
Syria, 96
Sytin, Ivan, 86, 87

Tahrir Square, Cairo, 68
Taiwan: and business and commerce, 72, 109; and Cable Law, 12, 63; cable television in, 11–12, 20, 55, 61–63, 72, 107, 109, 125, 127, 132, 133, 134; democracy in, 12, 61–63, 107, 129; government-controlled television in, 61–62, 134; government of, 132; liberalization of media in, 73, 125, 127; magazines in, 125; and national identity, 132; newspapers in, 125; political opposition in, 20, 21; politics in, 11–12, 61–63, 73
Taiwan Democratic Cable Television Association, 12, 62
Tarrow, Sidney, 33, 46–47
taxes, 89–91, 97, 98
Telecom Egypt, 69
telegraph: in Brazil, 8, 21, 77–80, 98, 130, 133; and Canadian Diamond Jubilee, 77; and Canadian national identity, 132; in China, 108, 129, 133; between countries, 79; expansion of resources through, 72; and governments, 74; international, 79; and trade and finance, 79; in United States, 82–83, 98

telephones: and Arab Spring, 124–25; in Canada, 21, 77, 80, 113, 115, 129, 132; in China, 132; and commerce, 80; and governments, 74; and International Campaign to Ban Landmines, 10, 20, 64, 73, 125
television, 109, 119–20; and Arab Spring, 124–25; audience of, 110; and Doordarshan's *Ramayana,* 91–93, 127, 132; government control of, 134; and governments, 74; in Philippines, 73, 125; and politics, 7; satellite, 21, 76, 134; and *TsunamiHelp,* 26, 48
texting. *See* Short Messaging Service
Thailand, 58, 88
Tilly, Charles, 114
Tolstoy, Leo, 86
Trans-Canada Telephone System, 77
trust, 13–14; and activism, 125; as binding communities together, 12–14; as built through work, 72; and communications technology, 5, 22; and communities, 16; and institutions, 16; and interaction, 13–14, 48, 104, 117; and Internet, 103; and networks as communities, 3–4; and reciprocity, 14, 103, 104, 105; in sharing stories, 117; and social capital, 22, 103; and technology, 104, 108, 116–17; through interaction, 99; and *TsunamiHelp* volunteers, 26, 38, 39, 43, 46, 47–48
tsar, 86–87
tsunami, of 2004, 9–10, 23, 28, 88
TsunamiHelp blog and wiki, 23–51, 116, 133–34; and collective participation, 42–46; cost of, 37; creation of, 23, 25, 32–38; and diversity, 111–12; effectiveness of, 26, 40–42, 43, 45; and information, 9–10, 23, 24, 25, 26, 28, 30, 31, 32, 33, 34–36, 37, 41–42, 43, 46, 47, 49, 50, 72, 107, 111, 124; and Internet, 20; and spammer attacks, 40; as transforming experience, 42–46
tsunamihelp.blogspot.com, 28
TsunamiHelp volunteers, 25, 41, 46–50, 107, 124, 128; and *blogspot.com,* 129; commonalities and differences among, 118; creation of blog and wiki by, 23, 25, 32–38; decision making by, 39; diversity of, 9–10, 111–12; effects of activity on, 26; egalitarian ethos of, 38–40, 49–50; and failure of institutions, 24, 131; and Hurricane Katrina, 44–46; initial reactions and motivations of, 28–29, 30–32; institution building by, 50; later activism of, 50; leaders of, 26, 37–39, 40, 45, 46, 48, 50; norms and routines of, 47–48; research on, 26–28; and social capital, 26, 46, 49; subsequent activities of, 26, 42–46
Tunicell, 69
Tunisia: cell phones in, 69; civil society in, 71; corrupt government in, 127; government of, 67, 68, 69; Internet in, 69, 70; protests in, 20, 55, 67–71, 72, 106–7; revolution in, 96; social media in, 68, 69–70, 72, 106–7
Tunisie Telecom, 69
Turnbull, Susan, 32
Twitter, 55, 67, 68–69, 71, 129
TXTGMA, 61
Tyler, Tom, 13, 107

United Kingdom, 24; and BBC, 93–95, 96, 97, 98, 111, 113, 127–28, 135; and Hutton Inquiry, 95; and South America, 79
United Nations, 9, 23, 42
United Nations Development Program (UNDP), 31
United States, 24, 58; and Canada, 77; commerce in, 83, 85, 98, 129; Federal Emergency Management

Agency, 45; and Hurricane Katrina, 88; Internal Revenue Service, 89–91; Internet development in, 83–84, 98, 129; and networking technology, 133; telegraph in, 82–83, 98
United States military: and Civil War, 82–83, 98; and computers, 85; and Internet, 83–84, 98

videocassette recorders, 11, 61, 62
videotapes, 73, 125
Vietnam Veterans of America Foundation (VVAF), 64
Viswanathan, Neha, 36, 41, 44
Voice over Internet Protocol (VoIP), 48
volunteerpoweronlineblogspot.com, 27

Walker, James, 105, 106
Wareham, Mary, 66
Wendy's (fast food chain), 93
Wikileaks, 71
Wikinews, 36–37, 124
Wikipedia, 36, 70–71
wikis, 9, 27
Williams, Jody, 10, 64–65
World Bank, 80–81
World Cup, 116
World Health Organization (WHO), 89, 91
World Trade Organization, 112
World War I, 87

Yew, Lee Kwan, 60
Youth for Change movement, 71
YouTube, 68, 69, 70–71